Civil engineering procedure

5th edition

'The art of directing the great sources of power in Nature for the use and convenience of man.'

The Institution of Civil Engineers

Civil engineering procedure

5th edition

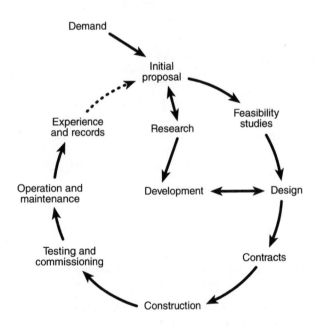

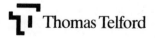

Published by Thomas Telford Publishing, Thomas Telford Services Ltd, 1 Heron Quay, London E14 4JD

First published 1996
Reprinted 1998, 2000, 2002

USA: American Society of Civil Engineers, Publications Sales Department, 345 East 47th Street, New York, NY 10017-2398
Japan: Maruzen Co. Ltd, Book Department, 3–10 Nihonbashi 2-chome, Chuo-ku, Tokyo 103
Australia: DA Books and Journals, 648 Whitehorse Road, Mitcham 3132, Victoria

Cartoons courtesy of Pat Atkins and Adele Stach-Kevitz
Figure 9 on page 52 appears by courtesy of Dr Martin Barnes

A catalogue record for this book is available from the British Library

ISBN: 0 7277 2052 X

Typeset by Selwood Systems, Midsomer Norton
Printed in Great Britain by The Cromwell Press, Trowbridge, Wiltshire

Preface

Civil engineering is 'the art of directing the great sources of power in Nature for the use and convenience of man' (Thomas Tredgold, 1788–1829). Its products are structures and systems for society, public services, commerce and industry. Civil engineering is thus central to the economy of the UK and other countries. It is also a major employer. The success of its projects is therefore vital to individuals, companies and governments.

The construction industry in the UK has changed and must change further to meet the demands of promoters for better value for money and to comply with greater legal regulation of construction health, safety and effects on the environment. The industry also needs to improve to eliminate late completion of work and disputes over extra costs.*

Traditionally the promoter of a project employed a consulting engineer to design the project and then employed a separate contractor to construct it, supervised by the consulting engineer. Alternative arrangements are now increasingly used for the design and construction of projects large and small, and also for maintenance, demolition and improvement work.

This guide is an introduction to the traditional procedure in civil engineering and to the alternatives. Chapter 1 describes the stages of work for a project, from the first consideration of ideas of what might be wanted through to the completion of construction and

---

* Sir Michael Latham, *Constructing the team*, final report, HM Stationery Office, London, 1994

'Are you sure it will last?'

handing over the resulting structure and services for use. Chapter 2 describes how a project may be initiated, and Chapter 3 the steps needed to define project objectives, investigate proposals and recommend whether to proceed further. Chapter 4 emphasizes the importance of determining a strategy for proceeding with a project and deciding who will be responsible for its design, construction and management.

The design stage, its management and the decisions to be made in design are described in Chapter 5, while Chapter 6 reviews the choices for employing contractors to construct a project. Chapter 7 describes the planning and control of construction, and Chapter 8 the management organization for that major stage of a project.

Chapter 9 describes the subsequent stages of final testing, commissioning and handover of the completed work. Operation and maintenance are considered in Chapter 10. Finally, comments are made in Chapter 11 on the trends and what the future may demand.

Appendices A, B and C list details of model conditions of contract, sources of further information and other details. Appendix D gives a Glossary of words used in this guide. The first time the defined words are used in the text of this guide they appear in *italic* letters.

Contents

1

Projects

Project cycle

Projects vary in scale and complexity, novelty, urgency and duration, but typical of most of them is a cycle of work as indicated in Fig. 1. Each stage in the cycle is different in the nature, complexity and speed of activities and the type of resources employed.

The durations of the stages vary from project to project, with sometimes delay between one and the next. Fig. 1 shows the common sequence of these stages. It is not meant to show that one must be completed before the next is started. They can overlap, for instance for an urgent project.

Initial stages

The cycle shown in Fig. 1 starts with an initial proposal to meet a demand for the goods or services the project might produce. The proposal is usually based upon engineering ideas and on experience and records from previous projects, together with information from research indicating new possibilities. The relative importance of information from research, demand, experience and records depends upon the extent of novelty of the proposals and how far innovations will be required in its design, but all these sources of information are always relevant to some extent.

Feasibility studies

If the first ideas indicate that the project may be economically attractive, the cycle proceeds to what are commonly known as *feasibility studies* to investigate a possible design and estimate its costs. Several alternative schemes likely to meet the expected demand usually have to be considered. In emergencies, this stage is omitted. If the project is urgent, little time is spent in trying to optimize the proposal. More commonly, alternatives have to be evaluated in order to decide whether to proceed and how best to do so in order to achieve the *Promoter*'s* objectives within his budget.

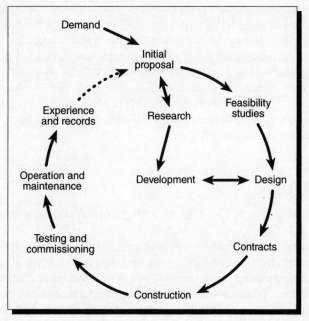

Fig. 1   Stages in project cycle

---

* The 'Promoter' is used in the rest of this guide to indicate the particular promoter of a project

'What if ...'

The results of the first evaluation may be disappointing. If so, the proposal has to be changed to try to meet the expected demand. The estimates of the demand may have to be reconsidered or made more precise, and the first evaluation may also have to be reviewed because information used has changed during this time. This stage may thus have to be repeated several times.

The investigation of schemes and their evaluation often proceed unsteadily because of the uncertainties in the information available for making decisions. The results are bound to remain uncertain. Estimates of the possible costs of the project depend upon the risks of construction and the future prices of materials, etc. Estimates of the potential value of the project depend upon the probabilities of the result meeting a demand when the time comes for it to do so.

Estimates of a project's potential value are particularly uncertain when a capital project which will not earn money is being considered, for instance a project to improve safety at a road junction. Nevertheless a specification, budget and programme must normally be decided, together with contingent margins of time and money, as the decisions made at this stage define the scope and standards of the project and should be the basis of all that is to follow.

Project selection

The investigations and feasibility studies of a proposed project may take time. The conclusions have to be quite specific: selection or rejection of the proposed project. This determines the project's future. Enough time and other resources should therefore be used to provide a valid basis for the decision. If the proposed project is rejected, it could be revived if new information is obtained on the demand, or a new design and other ideas are more economic. Otherwise it is dead. The information used for the feasibility studies should therefore have been good enough for that to be the best decision. Similarly, if a proposal is selected, the information used should have been good enough to provide the start to a successful project.

If the project is selected (*sanctioned*) the activities change from assessing whether it should proceed to deciding how best it should be realized and to specifying what needs to be done. Fig. 2 is a

common graphical way of showing the sequence of work for a project. This is an example of a bar chart in which the lengths of the bars indicate the expected durations of each activity.

Design

The decisions made early in its design determine almost entirely the quality, safety and cost, and therefore the success of a project. Design ideas are usually the start of possible projects. The main

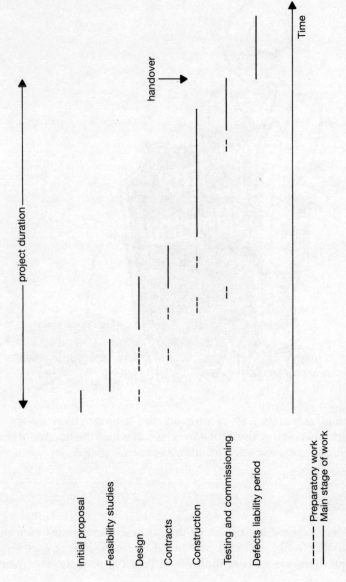

Fig. 2 Bar chart showing typical work sequence

design stage of deciding how to use materials to realize the project usually follows its selection, as indicated in Fig. 1. The products of design are usually drawings and a specification, but only a sketch may be needed for a small project, repairs or maintenance work.

Intermediate stages of design may be needed, for instance to provide sufficient detail to check estimates of costs and provide a scheme for approval by promoters and statutory authorities. For a novel project, further research and development work may be needed to investigate new or risky problems before the project is continued.

Contracts

Figures 1 and 2 indicate that a *contract* for construction follows the completion of design. This is normal in the traditional procedure for civil engineering and building projects in the UK. Alternatively, only an outline design or performance requirement can be the basis of a contract and the chosen *contractor* then becomes responsible for design and construction. *Consultants*, designers, *project managers*, suppliers and others are usually also employed under contracts,* some of them from the start of the investigation of a proposed project.

Construction

Construction usually requires larger numbers of people and a greater variety of activities than do the preceding stages. The cost per day rises sharply, as indicated in Fig. 3. So does the potential for waste and inefficiency. Construction therefore requires more detailed attention to its planning, organization, health, safety and

---

* Contracts to employ consultants are often called 'conditions of engagement' or 'service agreements'

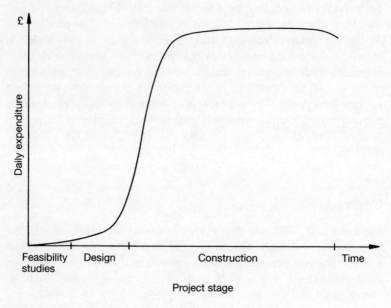

Fig. 3   *Cost per day against project stage*

costs. Demolition and substantial changes to existing structures require special care.

Most companies and public bodies who promote projects employ contractors from this stage on to carry out the physical work on site. Alternatively, contractors who take on the role of project promoters are normally fully responsible for design, for instance when investing in the construction of a building for sale or financing an infrastructure project such as the Channel Tunnel. For virtually all construction the contractors in turn employ *specialist* and local *subcontractors* to work on site and provide services, *plant*, materials and sub-systems.

Completion and handover

Sections of a project can proceed at different speeds in design and subsequent stages, but all must come together for commissioning and handing over the completed facility for use. The investment should then be achieving all the objectives of the project.

A subsequent proposal to alter the facility, replace its *equipment* and services, or decommission and demolish it is a separate project, and should proceed through an appropriate sequence of investigations and decisions.

Objectives

Each stage in the cycle shown in Fig. 1 should be planned and managed so as to provide the best basis for all the following stages. The purpose of the whole sequence should be to produce a successful project. The following chapters of this guide describe how each stage of a project can be organized to achieve this result.

2

Promotion of a project

Project promoters

Legal status of promoters

The promoters of civil engineering projects vary widely in their legal status. They include central, regional and local government and their agencies, limited liability companies, partnerships and businessmen who are sole traders. Other legally incorporated or unincorporated bodies may be promoters. Contractors are promoters of some projects, for instance toll road projects, power stations and property development.

Central government

Consultancy, construction and any other work done under contracts with UK government departments and agencies is paid for out of funds provided by a vote of Parliament. Expenditure in excess of the voted amount, although it may be authorized by a department or agency, needs a supplementary vote. The government then authorizes its departments and agencies to proceed with projects and studies for possible projects.

Contractors are not under any obligation to enquire whether or not the department or agency is contracting beyond the funds voted by Parliament. The government is bound by a contract made on its behalf by any of its departments or other agents which have the authority to enter into that contract.

Local government

The powers of local authorities in the UK are stated in their Charters,

Acts of Parliament constituting them and under general or special Acts governing their procedures and applications of funds. They can enter into contracts and raise funds for payments due under them. A contract validly entered into carries an implied undertaking that the authority possesses or will collect the requisite funds.

Local authorities are subject to the ordinary legal *liabilities* as to their powers to contract and their liability to be sued. Lack of funds is no defence to a legal action for payment. For example, a local authority's liability in a contract to pay for work will not be cancelled by the refusal of a government department to provide a grant or to authorize the expenditure.

Statutory boards and corporations

The powers of statutory boards and corporations are stated in the Acts of Parliament constituting them. The legal position of such bodies and of their officials is similar to that of local authorities. The former public utilities and nationalized industries which have been privatized are now incorporated companies.

Incorporated companies

Under English law an incorporated company can enter into contracts within the purposes of its memorandum of association or within the powers prescribed by any special Act of Parliament incorporating the company or any other Act granting it powers for a specific purpose.

Other bodies

A person or persons who are not a trading corporation, for instance a partnership or a club committee, can be the Promoter of a project. A consultant or contractor invited to enter into a contract with such a Promoter should check the authority of one or more of the individuals involved to commit the rest in personal liability for the purpose of that contract and for making payments due under it. The safest course for other parties is to enter into a contract with a sufficient number of members of the body to ensure that their collective financial status is adequate.

Overseas

The legal status of companies and of government bodies and the legal controls of construction vary from country to country, even within the Commonwealth or the European Union. Ascertaining the status and authority of promoters and others overseas needs local expertise.[1] Experience and advice are also usually needed on the culture and customs of construction organizations and individuals.

The World Bank, United Nations Industrial Development Organization, Asian Development Bank, European Investment Bank, European Bank for Reconstruction & Development and other funding agencies provide the finance for some civil engineering projects. Finance is also made available directly to poorer countries by the more wealthy countries, usually through government departments, for example the Overseas Development Administration, London, and development authorities in the country where a project is to be built.

Study team

Many industrial companies and government bodies employ their own engineering and other professional staff on the first studies and evaluation of projects. They also employ consultants, as do most smaller or occasional promoters. The Promoter usually enters into a contract (service agreement) for the consultant's services.

Consulting engineer

In the traditional procedure in civil engineering in the UK the Promoter employs a *consulting engineer* to investigate and report on a proposed project. The consulting engineer may be a firm or an individual, depending on the size and risks of the project. The consulting engineer's role at this stage of a project is to provide engineering advice to enable the Promoter to assess its feasibility and the relative merits of various alternative schemes to meet his requirements. Other specialist consultants may be needed, in a team led by the consulting engineer.

Project champion

Project sponsor

Many Promoters are organizations within which departments are employed on various functions such as the operation or maintenance of existing facilities, planning future needs, personnel management, finance, legal services and public relations. All these will have some interest in proposed projects, but they have different expertise and may have different objectives and priorities for each project.

A Promoter should therefore make one senior manager responsible for defining the objectives and priorities of a proposed project. It is the role of sponsor or 'project champion'. The role is not necessarily a separate one. It is logically part of the responsibilities of the manager who has the authority to ensure that sufficient resources are employed at this stage.

Project Manager

Projects are becoming more complex. Political, commercial, legal and technical change is such that specialists are needed to advise on the best way to develop and promote particular types of project. In addition, the general public will expect to be consulted on major projects or projects which may affect the environment and quality of local life.

In these circumstances the Promoter's interests may be best served by the appointment of one person to plan and manage the project and coordinate relationships with other organizations. This role is particularly important for

- ensuring that the project objectives are drafted for agreement by the Promoter and relevant financial or statutory authorities
- obtaining advice on the likely cost of the project, and possible sources of finance
- planning for site selection and acquisition
- planning for public consultation, a planning application and representation at public inquiries
- preparing the project strategy or 'brief' and planning for the appointment of the larger team and the systems, etc., needed for the next stages of the project.

Once a project has been sanctioned this person needs the authority to control its design and construction. In this guide the role is therefore called the *Project Manager*. Titles vary, as in so many jobs; what matters is the role.

The role of Project Manager is also not necessarily a separate job from other work, depending upon the size, risks and importance of a project. Because of the time it demands it is usually separate from the role of the senior manager who is project sponsor. The Project Manager may be the consulting engineer appointed to investigate and report on the proposed project. Alternatively the Project Manager may be an employee of the Promoter, or a specialist in project management.

The title 'Project Manager' is also used by contractors for their manager responsible for their work for a project. To avoid confusion in the remainder of this guide we use it to mean only the Promoter's Project Manager.

Selection of the team

Selection of a Consulting Engineer
The selection of a consulting engineer should start with defining the expertise appropriate to the project. Corporate membership of the Institution of Civil Engineers is recognized as the appropriate qualification for positions of responsibility in civil engineering. Corporate members of the Institution comprise Members, who have appropriate education, training and experience, and Fellows, who are senior members who have held positions of major responsibility on important engineering work for some years. All are subject to the Institution's by-laws, regulations and rules of professional conduct. Consulting engineers traditionally practised in partnerships, but increasingly these operate as limited liability companies.

Advice for Promoters regarding suitable consulting engineers and their methods of engagement and working can be obtained from the Association of Consulting Engineers. A formal agreement (a contract) should be completed between Promoter and consulting engineer which sets out the duties and responsibilities of each party and the fees and expenses to be paid. The terms of employment of a consulting engineer need to be consistent with those of others employed on a project, for instance the Project Manager and contractors. Sets of model terms are available for the employment of most professions.*

Overseas
Engineers working overseas must be prepared to adapt to different

---

* Model terms of engagement of consulting engineers and also project managers are listed in Appendix A

customs, contract arrangements, professional standards and terms of employment. In some overseas countries an engineer may practise only if registered for that purpose in accordance with the laws of the country. Some governments specify that any foreign consulting engineer must form an association or partnership with a local firm or agent before being permitted to practise in their country.

Selection of a Project Manager

The Project Manager should be appropriately qualified and have adequate experience of the type of project and the duties and responsibilities that the Promoter intends to assign to the Project Manager. For instance, a Project Manager for a civil engineering project usually needs to have the expertise to advise on risk management, project planning, contracts and organizing design and construction. The individual may also need the financial expertise to advise the Promoter on the expected whole-life cost of the project, cash flow and alternative sources of finance. An understanding of other professions may therefore be important for the Project Manager, particularly during the process of setting up a project team to manage a large and complex project.

The selection of the Project Manager should follow closely the process outlined for the selection of a consulting engineer. The Promoter can obtain advice on this from the Association of Consulting Engineers and the Association of Project Managers. Increasingly the promoters of civil engineering and other projects are employing Project Managers whose competence has been certified by the Association of Project Managers or the equivalent organizations in other countries.

Selection of project team

The Project Manager of a large or complex project will require engineering and other assistance. It is normal practice for the Project Manager to select and appoint the project team, but in some instances the Promoter may appoint the team on the advice of the Project Manager.

The selection and appointment of the team should be based on the same criteria as those applied in the selection of the Project Manager, care being taken to appoint a team that will have the expertise and the resources needed for the project.

Preparation of brief

Responsibilities
An important task for the Project Manager is to ensure that the Promoter defines the objectives for the project and agrees a project brief to guide the next stage of work. The brief should state

- the Promoter's objectives and priorities
- how consultants and other resources are to be employed
- an outline programme and budget, setting dates and cost targets for the investigations and design studies needed for the feasibility study.

The brief should be designed to guide the investigation and evaluation of alternative engineering schemes which appear on initial consideration to meet the Promoter's needs. An example of this would be a brief to investigate the provision of a bypass to a congested town. There may be several acceptable alternative routes and often within these routes alternatives such as a tunnel or bridge crossing to a river. For such a project, cost-benefit studies, risk analyses and environmental impact assessments for each alternative will help to concentrate the choice down to two or three alternatives which meet the criteria set by the Promoter.

Some investigative work may be needed to reach a decision on which of the alternatives merit further evaluation. Often a short study backed up by site visits and the use of already existing geological, hydrological and other information will be sufficient, together with the consulting engineer's experience on similar projects in the past.

Outline programme and budget
If a date for completion of the project has been set by the Promoter,

the outline programme should show a time limit for reporting back and for consequent decisions by the Promoter, dates for starting design, placing contracts and starting construction. The programme should also show any dates critical for financial and statutory approvals and agreements by others.

The basis is now established for the next stage, the investigation of the proposed project and reporting the results and recommendations to the Promoter.

Reference

1. Loraine R. K. *Construction management in developing countries*, Thomas Telford, London, 1991

3

Feasibility studies

Establishing the demand

The feasibility study of a proposed project should usually start with an investigation to confirm the expected demand for the goods or services that will be produced by the project. This may depend upon predicting the useful life of the project and the anticipated growth in demand during that useful life. Predictions of growth can be based on long-range economic forecasts, but in some cases uncertainty is such that the period of growth considered is arbitrarily limited to say 15 or 20 years.

Financial objectives

Normally the Promoter and others investing in a project will have in mind clear financial objectives for the project. These objectives usually include achieving a satisfactory cash return or a maximum total cost of the project.

A commercial project once completed is expected to earn money, i.e. to have a positive cash flow. 'Shadow' estimates of money values have to be used to represent the benefits of a project to provide services which do not earn money. In both the private and the public sectors it is normal for the Promoter to make a decision to proceed with a project based on calculations of the net present value and discounted cash flow over the planned life of the project.[1] Past projects, public data, risk analysis and estimating can usually provide a sufficiently accurate basis for predicting the costs of design, construction, operation, maintenance, decommissioning and demolition.[2] Shadow values can be difficult to agree, for instance some environmental costs and advantages, and therefore may be the subject of opinions and arguments.

The expected cash flow is usually important in planning and managing the design and construction of a project. Cost over-runs can ruin its viability. So can extra interest costs and delayed income if the project is completed late. On the other hand, completion ahead of programme can be embarrassing, depending upon the Promoter's arrangements to finance payment and contractual commitment to pay for work done.

Service objectives

The 'service' objectives for a project should define the capacity or performance required of it. They should also include requirements for the reliability of systems and equipment, freedom from lengthy downtime for maintenance, as for instance in the process industry, through to situations in which the finished project may be available for maintenance at regular intervals because of fluctuations in demand. An example of this would be a motorway, where the amount of traffic may be small enough at night or weekends to allow for access ('possession') for maintenance and improvement work.

The required life of the finished project is another objective which will have implications for design and construction. For instance, the economic life of a manufacturing project in an industry where technology or markets are changing rapidly is much shorter than that of a bridge which is useful for many years and produces no cash income by its replacement. Design to suit alteration or demolition will be important for the former, quality and minimum maintenance for the latter.

Statutory approvals

The execution of most projects in the UK is dependent on obtaining planning permission and other statutory approvals. Usually the Promoter has to obtain the statutory approvals, either directly or by appointing a consultant as his agent for the purpose.

Statutory Regulations demand compliance with specific procedures as laid down in the Regulations. The Promoter's team has to initiate a project in good time to submit the information needed to satisfy statutory authorities. Lack of experience and appreciation of the time required to get approval have been the cause of rejection of some projects, or of much additional expense which could have been avoided.

An early environmental impact assessment and construction health and safety study of all but the smallest projects, whether or not required by law, should enable decisions to be taken to anticipate serious problems before detailed design starts, thus avoiding the high extra cost of making changes later on.

Production of an environmental impact study, proposals for avoiding or mitigating serious environmental problems and the provision of positive environmental features will help gain approval of a project in public consultations, planning applications or a public inquiry.

Scope of investigations

The extent, nature and detailed content of the investigation of a proposed project should vary according to its value, urgency and complexity, the number of alternative schemes to be considered, and the nature and number of the decision-making processes involved before the recommended project can be constructed. For some projects, a single study with the preferred scheme and its estimated cost outlined in the report will suffice. Others will require a series of separate studies, each more detailed than the last.

The investigations for a major project may include some or all of the following

- outline design
- studies of novel requirements and risks
- public consultation
- geotechnical study of the site, sources of materials, storage areas and access routes
- environmental impact analysis
- health and safety studies
- testing for contaminated land and requirements for the disposal of wastes
- estimates of capital and operating costs
- a master programme of work, expenditure and financing
- assessment of funding.

The validity of the results depends upon

- the quality of the information used
- the novelty of the scheme
- the time available
- the amount and quality of the resources available to investigate the risks which could affect the project and its useful life.

These investigations require a range of engineering and other expertise. The Project Manager should ensure that every person employed on the work is briefed on the objectives of the project and that their tasks are planned and coordinated.

Information

Economic and safe design and construction depend upon information about the site. Ground investigation costs only a fraction of the total cost of the project. Adequate work at this stage will save money later on in the life of the project.[3]

Investigations should be carried out by experts.

All available data should be used, including existing topographical and geological maps and data, previous studies, aerial

'My design saves the cost of a site investigation ...'

and other topographical maps and marine charts. Data must be collected on the climate and the incidence of storms and flooding predicted. Additional data are usually needed in order to prepare an accurate plan and assessment of a site.

Items that may require investigation are

- conditions of access, for permanent use and also during construction
- environmental or other restrictions
- existing noise levels, to predict and be able to restrict as necessary, the noise levels during and after construction, including the combined effects of noise due to the project

> and to work on other projects proceeding at the same time
> - special rights of adjacent landowners
> - availability of services such as drainage, sewerage, water, electricity and gas
> - an archaeological study, as this may show the need for a full archaeological survey
> - the need for permanent or temporary disposal off-site of controlled waste, as the choice may be subject to planning approval. It may be advisable to consider securing options on soil disposal sites before seeking *tenders* for the project although this is traditionally the Contractor's responsibility.

Acceptable proposals on all of these can be important during public consultation.

Public consultation and participation

Most civil engineering projects affect the general public in some way. Although the public benefit from many publicly funded projects, there are usually some people who suffer. Examples of those suffering loss are the landowner whose land is compulsorily acquired for a new road scheme, and the householder who experiences noise nuisance as a result of the construction of a new airport runway. They may both reasonably have objections to a proposed scheme.

Involvement of the public at the earliest possible stage of both public and private projects can reduce objections and sometimes eliminate the need for a public inquiry. It can also produce constructive ideas to the benefit of the project.

Recommended design

Evaluation of alternatives

At this stage several alternative schemes may look economically attractive. The range of possibilities may vary between a number of solutions appropriate to one site only, to a number of solutions appropriate to several suitable sites. In the case of a new road project there may be several alternative routes, with the possibility of alternative detail within each route.

Alternatives can best be compared by considering the advantages and disadvantages of each systematically. The results should provide a ranked order of the options in terms of meeting the Promoter's objectives.

Their cost can provide the basis for comparison, but the conclusions can be uncertain because of the subjective nature of some factors such as those relating to the environment.* Constructing a matrix setting out scores for each alternative in terms of expected costs and cash flows, together with the values assigned to any more subjective factors under consideration, is a rigorous technique which can help avoid errors and also persuade others to agree the conclusions. If it is not possible to assign a cost to a factor then it is preferable to rank it against some ideal. This must be done for each option. In some cases the Promoter may require additional weighting be given to a particular factor or objective.

Design choices

All alternative schemes and the design choices which affect the economic viability of the project should normally be considered at this stage. The decisions can be flexible and alternatives can be considered relatively cheaply. Later changes incur the much greater costs of wasted work, wasted material, disruption of programmes and hasty work to recover time, as well as the cost of redesign. They can also affect motivation and the quality of work.

---

* The report *Assessing the environmental impact of road schemes*, Standing Advisory Committee on Trunk Road Assessment, HMSO, London, 1992, gives guidance which may be helpful on other civil engineering projects

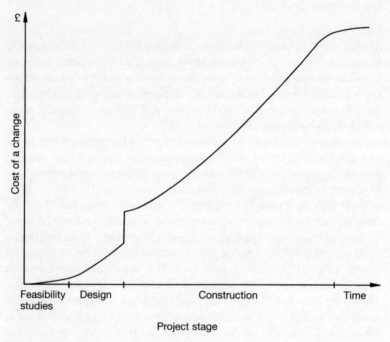

Fig. 4   Cost of change against project stage

Figure 4 indicates how the cost of a change can increase greatly during a project. The Project Manager should ensure that the Promoter and all concerned with the project understand this relationship and therefore make decisions on alternative schemes and design choices well before authorizing detailed design and preparations for construction.

The report

Objective
The Project Manager should present a report to the Promoter summarizing the investigations and the conclusions. Normally this

should describe the scheme best suited to meeting the Promoter's stated objectives and give forecasts of costs and financial viability to help the Promoter to understand the project as a whole.

Style of report

The style and form of the report should vary according to the Promoter's needs, the nature of the project and the purpose for which the project was commissioned.

The report should be written clearly and unambiguously. It should usually start with a non-technical summary so that it can be quickly understood and accepted by the Promoter and others who have neither the time nor expertise for detail. The body of the report should usually contain sufficient technical, programme and cost information to convince any other engineer or specialist that its analyses and recommendations are correctly based upon the information available.

Form of report

The report should start with a single page which summarizes what has been investigated, what are the recommendations, and what are the limitations due to time or information available. In most cases the main body of the report should include

- the Promoter's statement of objectives, priorities, initial instructions or terms of reference for the investigation
- statement of the need for the project, its importance, scope and history
- basic assumptions, data or trends on which the investigations have been based, and comments on their reliability
- review of the investigations carried out
- the design, health, safety, environmental and other criteria adopted and the reasons for their adoption
- a comparison of the alternative schemes considered and the reasons for the rejection of those not recommended

- cost estimates of the schemes, with notes on risks and accuracy
- a description of the recommended scheme
- a design brief to guide developing the scheme and detailed design
- programme for design, contracts, construction and handover of the completed project, with comments and recommendations on the time-critical and other risky decisions and actions needed to meet the objectives set for the project
- proposals for the project execution strategy, covering the organization and management of the design and construction of the recommended scheme, advice on the contract arrangements and, if necessary, the raising of finance.

References

1. CORRIE R. K. (ed.) *Project evaluation*, Thomas Telford, London, 1990
2. SMITH N. J. *Project cost estimating*, Thomas Telford, London, 1995
3. Site Investigation Steering Group. *Site investigation in construction series*, Institution of Civil Engineers, 1993

4

Project strategy

Scope and purpose

A recommendation to proceed with a project should include proposals for the organization and management of design, construction and other critical activities, as stated in Chapter 3.

If the recommendation to proceed with the project is accepted, the Project Manager should develop these proposals into a project strategy which states who is to do what and when. It should direct everyone to think about how the increasing scale of resources now needed should be organized, not just what has to be done.

Project management and control

Project Manager

From the start of design the Project Manager should be the only channel of communication between the Promoter and all the organizations or groups of people who will be employed by the Promoter to carry out the project. This is the lesson of many projects. The Project Manager should establish definitions of roles, authority, communication and reporting that bind everyone to the objectives.

Project team

Once a project is to go ahead a project team should be formed with the expertise and resources to assist the Project Manager to plan and manage all the remaining stages of the project through to handover. A contractor can be employed to design and construct from this stage.

Whether in the Promoter's or Contractor's organization, the team should include staff who have recent experience of the construction, operation and maintenance of similar projects. This should enable previous design weaknesses to be identified and so avoided, for instance problems of materials, processes or layouts which can affect performance or make maintenance tasks difficult or even impossible.

The structure of the team should vary for each stage of the work, and will depend upon the size and complexity of the project and how much it can draw upon the resources of established departments within the Promoter's organization, consultants and others.[1]

A rapid increase may be needed in the size of the team. Depending upon the project programme, some work can be undertaken in sequence or in parallel. Other parts of the work will be reliant upon information arising from earlier work and must be carried out sequentially.

Project controls

Project control depends upon information. Except for minor or emergency projects the Project Manager should establish procedures for

- work breakdown, definition of authority, responsibilities and control of changes
- planning and progress monitoring of office and site manpower needs, project design and services, preparation and placing of contracts and other procurement, construction, critical sub-contracts, testing, commissioning and handover
- health and safety plans, statutory approvals, audits and reporting
- cost estimating, cost monitoring and trend analysis
- equipment and material ordering, inspection and delivery
- quality management plans and audits

- project management information
- reporting to the Project Manager
- reporting to the Promoter
- document records and control.

Planning

The purpose of planning the work for a project is to think ahead about what is needed to achieve the objectives of the project. Programmes and reports should include the amount of detail needed by their intended users. Too little information can leave the team uncertain about what is wanted and what is happening. Too much can be counter-productive, as people will ignore it or at best only look for what they assume matters to them.

The planning of a project must allow time for the legal requirements to obtain approval of design and construction by statutory authorities. Planning techniques should be used which are appropriate to the scale, urgency and risks of the project.*

Monitoring and reporting

All but minor routine projects need a system for monitoring progress and costs from the start of design to provide a basis for regular reviews of achievement and trends compared to programme and budget. The scope of the data that are to be presented in a report should be agreed with its users.

Attention to cost trends and probable outcomes is important on most projects, the exceptions being urgent work. For all projects speed in reporting costs is usually more valuable than accuracy, to enable action to be take on trends. Quick data should be followed by accurate data and analysis of the causes of savings and extra costs in order to correct first impressions and improve the information available for estimating the costs of future projects.

---

* See the publications on project planning and control listed in Appendix B

Not everyone has the same view of a risk.

Risk analysis and management

Project risks
The information used to decide whether to proceed with a project is inevitably based upon predictions and assumptions about the future conditions and costs which may affect its design, construction and completion. Political events, weather, the quality of design, unknown ground conditions, bankruptcies, plant failures, industrial relations, accidents, mistakes and criminal actions are all risks which may affect the progress, cost or economic value of proceeding. The identification and assessment of possible risks is therefore valuable at every stage through a project.

Not every party to a project will have the same view of a risk and how it should be managed. The Project Manager should therefore establish the Promoter's policy on risks and inform all the project team.

Risk management
This is a general title for a systematic procedure for identifying, analysing and deciding responses to risks.[2] It consists of

Risk Identification. Experience and checklists are used to identify the sources of possible risks to the project, including physical, environmental, commercial, political, legal, financial, operational, technical, resourcing and logistical risks.
Risk Analysis. The probability and potential effects of each risk are assessed, first by judgement, and then quantitatively for the serious risks, using systematic procedures such as the 'hurdle' method, sensitivity analysis, decision trees, probabilistic analysis and simulation.
Risk Response. For each risk, decisions are made whether to

- ignore it, if it is too unlikely or its potential effects trivial
- eliminate it, by changing the project
- transfer it, usually to an insurer or a construction contractor

- bear it, allow for its possible cost and other effects and manage it.

Health and safety management

Attention to health and safety is a professional and legal duty of every person involved with the promotion, design, construction and operation of a civil engineering project.[3,4] The construction industry is one of the most dangerous. In addition to the human tragedies and waste, poor health and safety procedures at work and accidents affect morale and productivity and lose contractors and promoters money. Most of these could be prevented by planning, training, incentives and good supervision.

The most important pieces of UK legislation are the Factories Acts and the Health and Safety at Work, etc Act, 1974. Detailed requirements on sites are specified in the Construction Regulations. Requirements for attention to health and safety in design and construction are specified in the Construction (Design and Management) Regulations, 1994, (CDM). The legislation imposes criminal liabilities on everyone – promoters, consulting engineers, contractors, sub-contractors and every individual as well as organizations – to anticipate and control risks to health, safety and welfare as far as is reasonably practicable.

The Construction (Design and Management) (CDM) Regulations
The CDM Regulations apply to construction, demolition and dismantling work. Under the Regulations, designers and contractors have to be selected on the basis that they are aware of their health and safety duties and will allocate adequate resources to them.

The Regulations impose on the Promoter (called *the client* in these Regulations) the duty to select and appoint a *Planning Supervisor* and a *Principal Contractor* for a project.

The duties of the Planning Supervisor include

- ensuring that construction health and safety requirements are met in design

- ensuring that particulars of a project are notified to the health and safety authorities
- ensuring that a health and safety plan is prepared for the project and is finally handed over to the Promoter.

The duties of the Principal Contractor include

- maintaining the health and safety plan
- ensuring cooperation between all contractors on health and safety
- ensuring that everybody on-site complies with rules in the plan.

Selection of the Planning Supervisor
The person appointed must be competent and have adequate resources. The Promoter may appoint the Project Manager, a consultant or any other person as the Planning Supervisor. A member of the Promoter's staff may be appointed.

Selection of the Principal Contractor
The contractor appointed must be competent and have adequate resources. It can be the *main contractor* for constructing the project, a *management contractor* or a specialist in construction health and safety.

Quality management

Quality management is the process of ensuring that the project meets the Promoter's requirements for economic and safe quality of design, materials and construction.[5] Many promoters only employ consultants and contractors who meet national and international standards for quality assurance (QA). Training, quality assurance systems and detailed supervision on-site help greatly to achieve increased productivity and high standards of quality.

QA procedures are a systematic way of specifying actions which

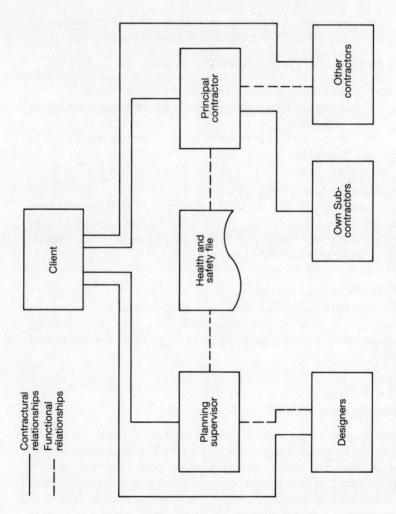

Fig. 5   Relationships of Planning Supervisor and Principal Contractor

will give confidence that quality requirements will be achieved. These procedures should

- state how the quality standards required for a project are to be decided and communicated
- establish how and how far proposed methods of work are to be assessed and improved to achieve adequate standards of work
- specify how the records of standards achieved are to be compiled, coded and recorded
- include auditing of the effectiveness of the procedures.

QA procedures should be clear and brief, their purpose being to anticipate problems. QA should be required for a project only if the cost of establishing and applying it is expected to be less than the total cost risks if there were no QA system. Quality management policies and QA systems should therefore be reviewed regularly to see if they are cost-effective.

Promoter's contract strategy

Contract strategy (or 'procurement strategy') is the process of deciding how to use consultants, contractors, suppliers and others, for the design, construction and also other services needed for a project. It should specify who should undertake all stages of work.

Strategy choices
A sequence of questions should be considered to choose a contract strategy:

1. Should the Promoter employ consultants and contractors, or carry out all or some of the work with own employees?
2. How many consultants and contractors to employ? If more than one organization, are they to be employed sequentially or together?
3. Who is to be responsible for what? Who is to be responsible

for defining objectives and priorities, design, quality, operating and maintenance decisions, health and safety studies, approvals, scheduling, procurement, construction, equipment installation, inspection, testing, commissioning and for managing each of these?

4. Who is to bear which risks? Who is to bear the risks of defining the project, investing in it, obtaining the necessary approvals, specifying performance, design risks, ground conditions, selecting sub-contractors, site productivity, mistakes and accidents?

5. What terms of payment will motivate all parties to achieve the Promoter's objectives – including payment for design, equipment, construction and services?

Some of the answers may be dictated by law, government policy or by financing bodies.

Contracts for construction work

It is usual to invite contractors to compete for a contract for construction work, in the expectation that they will plan to do the work efficiently and therefore at minimum cost. This is usually required for public works projects. They are invited to tender on an equal basis, competing on price and evidence of past performance. Competing on price and speed of starting work is common practice for sub-contracts.

Most contracts are for the construction of a defined project. Ideally the detail is complete and final when tenders are invited. If not, the contract terms allow for some *variations* to design and construction during the work.

The value of time to a Promoter may be so great as to dictate a start to the construction of a project before design is complete and final. This strategy is known as 'fast-track' construction (in the USA 'fast-trak'). The value of inviting competitive tenders may then be limited by the incompleteness of design. The *cost-reimbursable* or *target-cost* types of contract discussed later in this chapter are a possible basis for such contracts. These are more flexible, but require more detailed planning and control by the Promoter. They are used if the objectives of a project are subject to continuing change or uncertainty, or if its design depends upon new technology or has to await detail of mechanical and other systems and equipment.

On operating sites, for instance factories, contractors are also employed under *term contracts* for carrying out investigations, design, construction, maintenance or demolition work when required at any time during a year or longer term.

Contracts for consultancy services
Traditionally consulting engineers in the UK and internationally were employed by promoters under standard terms of engagement and fixed scales of fees stage-by-stage through a project. Other consultancy work for promoters or contractors was most often reimbursed at cost plus a fixed fee. Consultants are now increasingly invited to compete by price to provide design or other services, whether employed by promoters or contractors.

Contract responsibilities

Responsibilities and duties of the Promoter
The Promoter's objectives, responsibilities, duties and liabilities should normally be stated in all contracts with consultants and contractors, including

- defining the functions that the project is to perform
- providing information and data held by him and required by the other parties
- obtaining the necessary legal authority to allow construction of the project
- acquiring the necessary land
- making payment

The Promoter may of course arrange for some or all of these duties to be performed by a consultant or a contractor.

Traditional arrangements

Separation of design and construction roles evolved in civil engineering in Britain in the great age of road, canal and railway building which began in the eighteenth century. The services of civil engineers able to design novel projects successfully became in great demand, especially to advise promoters on the cost and the planning of large-scale investments.

There emerged separate firms of public works contractors who mobilized the people and the plant to construct the projects. Most of these became limited liability companies. The design engineers became consulting engineers independent of the financing of construction, and mostly in partnerships employing qualified and more specialized engineers. A similar separation of roles evolved between architects and building contractors. Consulting engineers established professional firms, to be engaged by promoters to study ideas for new projects and to report on costs, problems and how to proceed. If a Promoter decided to proceed, the consulting engineer was then usually appointed to undertake design, prepare specifications and other contract documents, and to advise on choosing a contractor. In the traditional arrangements for a public works project in the UK one consulting engineer is appointed as leader of the project team, and one main contractor employed for construction.

More recently many UK government authorities and industrial promoters have reduced their engineering staff and used an increasing variety of alternative contract arrangements such as employing contractors to design and construct all or most of a project. Former nationalized industries responsible for water, gas and electricity supply have tended to do the same when privatized.

The Engineer

Traditionally in civil engineering in the UK the consulting engineer is the adviser to the Promoter from the inception of the project, as described in Chapter 2. When the project goes ahead the consulting engineer is appointed to be responsible for design, preparation of the construction contract and tender assessment. The Promoter also appoints the consulting engineer to be *the Engineer* named in the construction contract with powers and duties to supervise *the Works* and make decisions on design, construction and payment.

In this system the Engineer (with assistants) administers the contract between the Promoter and *the Contractor*, as illustrated in Fig. 6. During the execution of the Works his duties include the inspection of materials and workmanship, monitoring progress, and a variety of administrative duties such as the *measurement* and *valuation* of work done and the pricing of varied work. The Promoter can appoint his Project Manager as the Engineer.

In some contracts the title 'the Project Manager' is now used to mean the role of supervising the Works. To avoid confusion, in this guide we use 'the Engineer' for this role. Other titles including 'the Architect', 'Supervising Officer' and 'Contractor Administrator' are also used for similar roles in large building and some civil engineering contracts. The person's powers and duties vary from contract to contract, but in principle operate as indicated in Fig. 6.

The role of the Engineer with powers and duties to supervise the Works is also used in international civil engineering contracts.

General Contractors
General contractors are those who, on account of their resources and experience, are able to undertake responsibility as main contractor for the construction of the whole of a project, although they may *sub-let* parts of the work to specialist or other sub-contractors.

Specialist contractors
Those who confine their activities to selected classes of work are referred to as specialist or *trades contractors*. This specialization enables them to employ skilled staff and plant particularly suited to their work. In some cases their designs and techniques are protected by patents.

A specialist contractor may perform his work by sub-contract to a general contractor who will act as coordinator. Alternatively the specialist contractor may perform his work under the direction and supervision of a management contractor or by direct contract with the Promoter.

The introduction of new designs, processes and methods of construction is often due to the activities of such contractors and their employment can be of economic advantage to both Promoters and general contractors.

Alternative contract management arrangements
A relatively new development for many UK promoters is appointing a contractor as management contractor to plan and supervise trades contractors (then called *works contractors*) to construct the project. Fig. 7 illustrates this arrangement. The management contractor is usually required to be responsible for the measurement

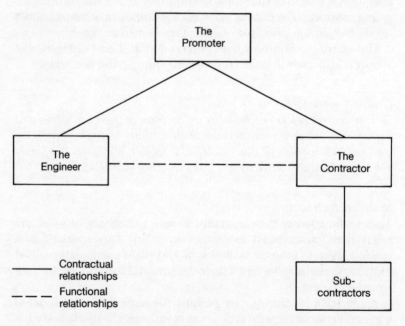

Fig. 6   Relationships of the Promoter, the Engineer and the Contractor

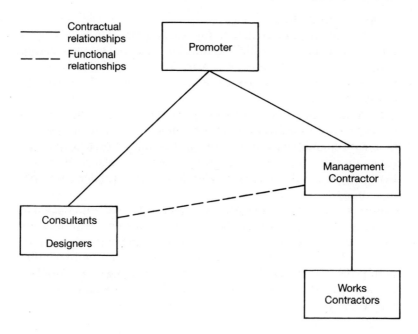

Fig. 7   Relationships between the Management Contractor and others

and certification of payment due to the trades contractors. A management contractor is usually paid a fee which may be partly a *fixed price* and partly dependent upon the performance of the trades contractors.

In this and alternatives known as the *construction management* and the *project management* systems the Promoter employs the managing organization as an extension to his own organization. Their duties vary from project to project, for instance in responsibilities for coordinating design and construction. The duties and powers of the managing organization need to be stated consistently in its contract with the Promoter and in the contracts of the trades contractors.

Comprehensive contracts

A rather different alternative is for one contractor to have a comprehensive contract in which he is responsible for complete design and construction. These contracts are known by various descriptions, including Design-and-Construct, Package-Deal, All-In, Turnkey.

In these contracts the contract stage shown in Fig. 1 comes before the detailed design. To undertake them some contractors employ their own engineering and architectural design teams. Most employ consulting engineers, architects and other specialists when needed for a project, for all the design work or to augment their own core of staff.

In this arrangement, construction can be planned with design of a project. If this option is chosen, the Promoter's requirements should be detailed and final before tenders are invited, to avoid changes later. Comprehensive contracts have become more popular with promoters in the UK, and it is also becoming more common for both UK and overseas contractors to have to obtain the finance for a project.

Concession contracts

Recent infrastructure developments in the UK and many countries have been constructed under Build-Own-Operate-Transfer (BOOT) or Design-Build-Finance-Operate (DBFO) concessions to contractors and others, often in joint ventures. In these, the comprehensive contract includes responsibility for financing and managing operations and maintenance for a fixed franchise period. During this period the concessionee receives the income from users, for instance from drivers using a toll road. At the end of the period, ownership of the project usually transfers to the government.

A variety of concession arrangements have been developed, with various names, but all have in common that contractors and their partners obtain the finance to construct the project and they are therefore the project promoters.

Project financing

How a project is to be financed may affect how or which consultants and contractors can be employed, for instance for projects in developing countries. For these, the choice of consultants or contractors is usually limited to those approved by a financing organization, and the choice of contractors to those who can finance construction.

International practice
Similarities to what has been called the British traditional arrangements are to be seen in Commonwealth countries and on projects in the developing countries financed by international banks. Variety of systems is more typical of the USA and Canada. Employing several separate specialist trades contractors in parallel, without a main contractor, is more typical of construction in other European countries.[6]

Sub-contracts

A common principle is that a main contractor is responsible to the Promoter for the performance of his sub-contractors. Practice varies in whether a main contractor is free to decide the terms of sub-contracts, choose the sub-contractors, accept their work and decide when to pay them. It also varies in whether and when a Promoter may bypass a main contractor and take over a sub-contract.

Civil engineering contracts normally state that the Contractor shall not place any sub-contracts without the approval of the Engineer, and that the Contractor shall remain liable for all the acts and defaults of sub-contractors.

Nominated sub-contractors
In some model *conditions of contract* a sub-contractor can be *nominated* by the Engineer. This is done for instance if the sub-contractor has to be ordered to start to manufacture a component for a project before the main contract is made, if a particular supplier's product is required, or if novel or risky work is required from a specialist sub-contractor under direct supervision by the Engineer. Similarly,

in some contracts for building work the Architect or equivalent can nominate sub-contractors.

In these arrangements the Contractor is instructed to obtain quotations from approved sub-contractors, and then accept the tender of the nominated sub-contractor and work with that sub-contractor as with any other. The Contractor usually has the right to decline to accept a nominated sub-contractor for a good reason.

The nomination of sub-contractors is not recommended, because it complicates relationships and divides responsibilities. Parallel contracts with the promoter is the alternative arrangement, as is usual in industrial construction.

Internal contracts

A Promoter may choose to have work constructed by his own maintenance or construction department, known in the UK as *direct labour* or direct works, instead of employing contractors. If so, in all but small organizations the design decisions and the consequent manufacturing, installation and construction work are usually the responsibility of different departments. To make their separate responsibilities clear, the order instructing work to be done may in effect be the equivalent of a contract that specifies the scope, standards and price of the work as if the departments were separate companies. Except that disputes between the departments would be managerial rather than legal problems, these internal 'contracts' can be similar to commercial agreements between organizations.

Many local government authorities in the UK have direct labour organizations (DLOs) which carry out at least small projects and maintenance work. Now compulsory competitive tendering (CCT) legislation requires their DLOs to compete for most work with independent contractors.

If a contractor promotes as well as carries out a project, he may need to separate these two roles because different expertise and responsibilities are involved in deciding whether to proceed with the project and then how to do it. Separation of these responsibilities may also be required because others are participating in financing the project. For all such projects except small ones, an internal contract may therefore be appropriate to define responsibilities and liabilities.

Project execution plan and procedures

The results of the decisions on all the above should be the proposed strategy for a project. The Project Manager should agree it with all the managers who control resources and obtain the Promoter's acceptance of the strategy before proceeding with the project.

The strategy requires publicizing to everyone working for the project, at least to the extent of stating the project objectives and scope. A small guidance note or a display specific to a project can achieve this. For larger or novel projects a 'project execution plan' or 'project implementation plan' and displays are required.

Whether or not a project needs a detailed separate plan or only a document that supplements established standards and procedures, every person who is to be responsible for planning and managing work for it should be told

- the Promoter's objectives

  - purpose of the project
  - performance criteria and constraints
  - quality and safety standards required
  - completion date, and any intermediate dates of importance to the Promoter
  - cost limits
  - the priorities between time, quality and cost

- the risk and safety management policies, and any special requirements or constraints
- the organization of the work – the work breakdown structure and the contract strategy

- the role and organization of the project team and supporting resources
- the system for project communications, control and management.

The execution plan should be used to guide all the work that follows.

References

1. WEARNE S. H. *Principles of engineering organization*, Thomas Telford, London, 2nd edition, 1993
2. THOMPSON P. A. and PERRY J. G. (eds) *Engineering construction risks – a guide to project risk analysis and risk management*, Thomas Telford, London, 1992
3. INSTITUTION OF CIVIL ENGINEERS. *Managing health and safety in civil engineering*, Thomas Telford, London, 1995
4. HEALTH AND SAFETY COMMISSION. *Guidance notes on the Construction (Design and Management) Regulations*, London, 1995
5. BADEN HELLARD R. *Total quality in construction projects*, Thomas Telford, London, 1993
6. EUROPEAN CONSTRUCTION INSTITUTE. *Construction contract arrangements in EU countries*, Loughborough, ECI, Loughborough 1993

5

Design

Design process

The aim of the designer should be to ensure that sufficient information is produced at each stage of the process to

- show that the project will achieve the Promoter's objectives
- obtain any necessary approvals and consents
- define in detail the next stages of the project, particularly the drawings and specifications needed for contracts, construction, testing and commissioning.

The successful design of a project demands not only expertise in technical detail but also a wide understanding of engineering principles, construction methods, costing, safety, health, legal and environmental requirements. In the case of a river crossing, for example, should the solution be over, under, around or on? What will be the cost, how can it be built, how long will it take, and what will the effects be on society and the environment? The resulting information should specify what is to be built, the standards and tests required, health and safety criteria, constraints on construction methods, and the life-long inspection and maintenance requirements.

As described in Chapter 2, design begins at the first investigation of ideas for possible projects. It becomes progressively more defined through the investigation of schemes and alternatives. It usually proceeds in stages as illustrated in Fig. 8.

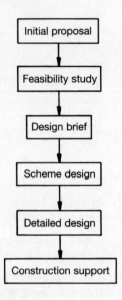

Fig. 8   Stages in development of project design

The design of a project can proceed through all these stages quickly if the project is small or similar to previous ones. Separate defined stages will be needed where the responsibility for design is to pass from one organization to another.

Design brief

For feasibility studies, the designer must advise the Project Manager on the alternative possible schemes which might meet the Promoter's needs, and their risks. The result of the feasibility studies should be a design 'brief' which defines the scope, objectives, priorities and design criteria of the proposed project.

The brief should provide the guidance needed to undertake the detailed design. Changes to it should be made only if the information used in the feasibility studies has changed. Changes up to this stage usually have only a small effect upon the whole project cost. As shown in Fig. 4, much greater costs are incurred if later changes require re-design, discarding of work done or delays to construction. Many considerations are likely to apply during the stages of the design process. Some of these are as follows.

Design programme
The design brief should include a schedule of dates for delivering drawings, specifications, calculations and other information for cost estimating, contracts, ordering materials and, of course, construction. If the project is complex, uncertain or urgent, every stage may need detailed planning and coordination with site investigations, risk studies, modelling work, the timing of submissions for approval by the Promoter and planning authorities, environmental impact studies and the health and safety plan.

The project programme should give time for the design team to visit the site to assess design and construction problems, and if needed to arrange for surveying, site investigation, traffic counts, flow measurements and further studies such as the location of supplies of materials, use of local services, temporary traffic needs and so on.

The designer should advise the Project Manager of the time needed to prepare designs, contract drawings and a specification, as lack of time for these is highly likely to lead to delays and additional expenditure.

Quality–cost–time
Of all the considerations, arguably the most important is the relationship between the quality and performance of the completed project, its cost and the time taken for the work, within the requirements of health and safety. This need to consider these together is illustrated in Fig. 9.

Directions on balancing the cost, quality and speed of construction should be given in the design brief, as the achievement of one can conflict with the achievement of the others.

Fig. 9   Relationship between quality, cost and time

Health and safety
The design must comply with the health and safety standards required by law. The main duties of the designer under the CDM Regulations are to

- alert the Project Manager to the Promoter's duties
- consider during design the hazards and risks which may arise to those who will construct and maintain the Works
- design to avoid risks to health and safety throughout the life of the project
- reduce risks at source if avoidance is not possible
- consider measures which will protect all workers if neither avoidance nor reduction to a safe level is possible
- ensure that drawings, specifications, operations and maintenance instructions, etc., include adequate information on health and safety
- pass this information to the Planning Supervisor so that it can be included in the health and safety plan
- co-operate with the Planning Supervisor, and where necessary other designers involved in the project.

Aesthetics
The Promoter's brief to the designer should state the policy on the aesthetic aspects of the design. Aesthetics should be considered by

the designer. For example, materials need not be visible in their base form. More attractive finishes can greatly improve the quality and durability of the finished product at little extra cost. Novel designs need to achieve a balance between whole-life cost and aesthetics.

Roles and organization

The brief should also state who is to be responsible for the remaining design work. Roles in the design team need to be defined, in particular who will lead the team.[1] Specialist design contributions may be necessary. Multi-disciplinary team working can be effective provided that there is no confusion of roles or duplication of effort.

Under the CDM Regulations, the Promoter has a duty to appoint designers who have the competence and resources needed for their duties under the Regulations. The design team may include members of the Promoter's permanent staff or external personnel, specialist firms or contractors employed for consultation and for design of sections of the project. The design of specialist work may be contracted to firms who later become nominated sub-contractors for the design and construction of parts of the project, but the Promoter's designer should retain ultimate responsibility for the design of the project as a whole.

The design team should be located where the main issues need to be resolved, and therefore on a complex project some or all of the team may have to move to the project site at the start of construction.

Outline design

Outline designs giving more information than used for the feasibility study may be required for approval by the Promoter and statutory authorities. If so, the designer should provide the Project Manager with the information required for the Promoter and for obtaining planning and other approvals and consents.

Planning permission may be granted in one or two stages; first for an 'outline' submission which may be accepted subject to conditions, and secondly for a detailed submission. Time and resources are needed in order to achieve this.

The designer may also have to prepare an Environmental Impact Analysis, with evidence that outside parties who may be affected by a project have been consulted and their interests considered. For this and other work which draws on specialist knowledge, the designer has to plan and manage interfaces with other professionals, the public and other third parties, in order to obtain information and agree a design which meets the Promoter's objectives and priorities. The Engineering Council provides engineers with a code of professional practice on environmental issues.[2] This includes guidance on the need to seek ways to change, improve and integrate designs, methods, operations, etc. to improve the environment.

Scheme design

Designs must be economic to construct. The availability of appropriate resources – materials, labour and plant – for construction is important to the economics of the design, particularly at geographically isolated sites.

Contractors' experience can be invaluable in achieving practical and effective cost and programme solutions. Construction expertise should be applied before detailed design, rather than alternative designs being received from contractors with their tenders just before construction. Traditionally in the UK the experience of contractors was brought in only at the tender stage. Consultation with contractors during the design stage is now preferred by many promoters to ensure that the design is suitable for economic and safe construction, but the consultations need to be conducted fairly between contractors who may later be competing for a contract to construct the project.

The extent of design information needed for inviting tenders depends on the method for procuring construction. The Project Manager should therefore ensure that the type of contract to be used for construction is decided in good time for the designer to be able to produce the type and extent of the documents required for inviting tenders.

If contractors are being invited to be responsible for detailed design as well as construction, the Promoter's designer should assist in their *pre-qualification*, tender analysis and selection. Only a performance specification may be needed for inviting tenders, but the Promoter will usually employ a design team to check the detail. In the traditional procedure, the detailed design of a project should be completed before tenders for construction are invited. In some cases, this is not practicable because of the urgency of the work, and the drawings on which tenders are invited are supplemented by further drawings issued by the designer during construction.

Some site or other information may be known only when construction is under way and in such cases, redesign or supplementary design work during the construction phase may be unavoidable. Typical reasons for this are

- excavation in ground which proves to be different from that inferred from site investigations
- structures to house equipment, the details of which are unknown at the design stage.

Whole-life requirements

The designer will normally need to design for economic performance during the whole life of a facility after its construction, through the stages shown in Fig. 10.

Use and maintenance
Design which uses standard materials and components makes replacements cheaper and easier to obtain. This can be important to the Promoter, not only to permit quick recovery in emergency situations but also to obtain more components for a later expansion of the facility. The costs of maintenance should be considered early

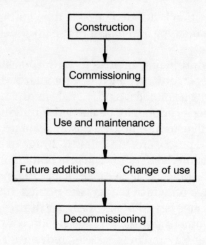

Fig. 10   Stages in the whole life of a facility

so that the design can incorporate details which may be maintained economically. If possible, experienced operating and maintenance staff should be consulted for their requirements for access to inspect, maintain and replace structures and equipment throughout their useful life.

Future additions and change of use
Before proceeding to detailed design, the designer should suggest to the Project Manager how the design could include provisions which would facilitate future extension or changes. These provisions can produce a project of much greater value to the Promoter at little extra initial cost.

Decommissioning
Designs have finite lives and at some stage systems and structures have to be decommissioned and demolished. The designer can prepare for this by, for example, ensuring that structural frames are

easily identifiable, specifying construction procedures that may be easily reversed, specifying materials that can be recycled or that generate low quantities of waste, and incorporating quality features into equipment design to safeguard against spillages of potential contaminants during operation.

Detailed design

Detailed drawings and calculations are prepared for two purposes – statutory approvals and construction. Unless the project is urgent, the completed detailed design should also be the basis of a re-estimate of the cost of the project before being committed to a contract for construction. This re-estimate can be used to judge tenderers' prices.

Once the detailed design has started, changes should be allowed only if essential for the satisfactory completion of the project. A change control procedure should be applied so that the total potential costs and time effects of any proposed change may be determined.

The designer should be expert on specifying materials, recognizing that ongoing research and development regularly yields new products. The engineering properties of materials should be understood and development work should be undertaken if needed to test new materials and methods.

Standard and locally available materials and components should be chosen wherever suitable, in order to reduce construction costs, make planning times more reliable and save training of construction employees. Innovations in construction methods and materials should therefore be considered early in design, in time to investigate their advantages to the Promoter's objectives. At the detailed stage it is wise to use proven technology, methods and materials for all time-critical aspects of a project.

Design methods

Calculation and analysis

Design is usually carried out in at least two main stages. For a building structure, the first stage is to calculate the loads approximately with an analysis of the proposed design. When the analysis is checked and found satisfactory the loads are then calculated more accurately. In this, established standards and guides should be used wherever appropriate, to achieve reliable results economically.

The second design stage consists of proportioning the components of the structure according to the results of the calculations, together with an adjustment of the original calculations to any altered sizes of parts. The staged process is usual in the design of all types of project. Design iterations are often required to adjust parts of the design to detail from the designers of systems, plant and services.

Checking and assessment

All design should be checked, and the checking formally recorded on every drawing, specification and set of calculations. The assumptions and methodology employed in calculations should be subject to a review of principles. Assessment of all design by an independent party is recommended, and is statutorily required for larger structures.

Value engineering

Value engineering is an analytical technique for questioning whether the scope of a design and the quality of the proposed materials will achieve the project's objectives at minimum cost.[3] It can be used at every stage of design to check that the Promoter will get what he needs. It gives an opportunity for the designer to prove that design choices are economic and to identify where costs can be reduced.

Review and audit

Depending upon the project strategy on risks, the design process may be also monitored by independent review and audit. The application of a quality assurance policy may require audits during any stage of the design process. The designer needs to allow time for these audits and for any repetition of the work shown to be necessary.

Information technology
There are many computer-based tools to assist the designer to prepare and revise the design quickly and efficiently. Computer-aided draughting (CAD) is now used for studying alternative layouts, design coordination, checking clearances for construction, operations and maintenance, presenting the design, taking off quantities and revising detail.

The designer must understand how the computer packages employed operate, in order to be confident that the results are accurate. The designer must not lose the skill to apply engineering principles to determine solutions, and should seek to validate the operation of packages used and verify computed solutions.

Consultation and approvals
The designer may need to present the design to the Promoter and representatives of statutory authorities, the public and others for their comments and approval. To get their approval the designer will need to understand the nature of objections to proposals and gain acceptance of the final design. On a large or novel project this may have to be done several times as the design is being developed.

The designer needs to develop good presentation skills and recognize the value of good presentation material. Time and training may be needed to develop these. The timing of presentations and consultations can be critical to the success of the project programme.

Construction support

The designer is usually required to support site staff by

- clarifying the design and offering redesign, especially if unforeseen conditions are encountered
- checking and approving design carried out by the Contractor or sub-contractors, at least for *temporary works* which are the Contractor's responsibility
- specifying tests and assessing the results
- assisting in producing the as-built drawings.

The designer should therefore not only inspect the site thoroughly at the start of his work, but also visit it regularly during construction and be available until the project is complete and handed over.

References

1. RUTTER P. A. and MARTIN A. S. *Management of design offices*, Thomas Telford, London, 1990
2. ENGINEERING COUNCIL, LLOYD'S REGISTER and DEPARTMENT OF THE ENVIRONMENT. *Professional practice, engineers and the environment*, London, 1994
3. INSTITUTION OF CIVIL ENGINEERS. *Value engineering*, Thomas Telford, London, 1995

6

Contracts for construction

Successful contracts

A contract is an agreement which is enforceable at law. It commits all the parties to obligations and liabilities. Typically a contract for construction commits the Contractor to construct the Works and the Promoter to pay. Both are obliged to endeavour to make it possible for the other to carry out their commitments. A breach of the contract by either party may make that party liable for paying damages to the other.

A Promoter will usually wish to know as accurately as possible before entering into a contract for the construction of a project when the work will be completed and the total cost he may have to pay. In most cases the Promoter will also wish to ensure that the work is done at the minimum cost that achieves satisfactory standards of construction health, safety and quality. The extent that these objectives can be attained depends largely on

- the quality and completeness of the information on the project and conditions for the work which forms the basis of a contract
- the control of changes to that information
- the choice of *terms of contract* appropriate to the uncertainties and the risks.

The same applies to contracts for site investigation and demolition work. The more that the information for a contract is complete and final, the better all parties will be able to forecast the completion date and final cost of a project.[1] The more that the Promoter accepts the appropriate allocation of risks, the greater the chances of achieving these forecasts.

Contract documents

The contract documents must specify the scope, location, quality and type of work to be carried out. A brief description of what is to be done may be sufficient for a contract for small and familiar work. Many drawings, a specification and CAD data may be required for large or novel construction.

Also important are what are called 'conditions of contract' or 'general conditions of contract'. These usually cover

- who is responsible for design, construction and supporting work
- how risks are shared between Promoter and Contractor
- procedures between the parties, particularly arrangements for monitoring and controlling quality of work
- use of sub-contractors
- programme of work, time extensions, completion, take-over
- actions on delay, defects and other failures to perform
- terms of payment
- insurances
- security for performance
- variations
- termination of contract
- settlement of disputes.

Model and standard conditions of contract

Model or standard sets of conditions of contract have been published by the UK Engineering Institutions and other bodies. They vary in their terms to suit the needs of different projects, particularly in who is responsible for design and in the terms of payment. Use

'All communications should be expressed in writing'

of the appropriate set of conditions without additions or alterations is recommended. Here we briefly describe the principal ones used for contracts for the construction of civil engineering projects. Their full titles are listed in Appendix A.

The ICE 6th edition Conditions of Contract (ICE Conditions or ICE 6)
The ICE Conditions have been developed for what is called the traditional civil engineering procedure for construction. They provide for detailed design to be done by the Promoter's design team and construction by the Contractor. Payment to the Contractor is made monthly, on the basis of *admeasurement* of the quantities of work done at tendered *rates* for the items of work listed in a *bill of quantities*. As in other civil engineering conditions of contract, the Promoter is called *the Employer*. The Engineer has the distinctive

duties in the contract of operating impartially and also as the agent of the Promoter.

A shorter version known as ICE Conditions of Contract for Minor Works has been published (see Appendix A).

ICE Conditions for Design and Construct have been developed from the ICE Conditions to provide for the Contractor to carry out design to the 'Employer's Requirements' as well as construction of the Works (see Appendix A).

ICE Conditions of Contract for Ground Investigation have been produced for site investigation work.

The Federation of Civil Engineering Contractors (FCEC) has published sub-contract conditions for use with the ICE Conditions, known as the 'blue' conditions.

The ICE New Engineering Contract family (NEC)
Because of the rapidly changing situation in contract systems and much dissatisfaction with current forms of contract and their proliferation, the Institution of Civil Engineers has produced the radically new New Engineering Contract conditions of contract. Its objectives are

- flexibility
- clarity
- stimulus to good management
- avoidance of confrontation and disputes.

The NEC consists of a family of conditions of contract (listed in Appendix A). The NEC is designed for use on civil, process, electrical, mechanical and building work or any combination of these disciplines, and for any proportion of design by the Contractor. By combining core clauses with clauses in one of the main options, a variety of payment methods are possible including those required for priced contracts, target, cost reimbursable and management contracts. The choice of clauses from a range of secondary options makes possible a considerable variety in the sharing of risks between the Promoter and Contractor. A new option is a clause that establishes a trust fund to guarantee payment if a party in the supply chain becomes insolvent.

The NEC incorporates management procedures in which the parties participate. This is designed to motivate the parties to cooperate, and thus reduce confrontation and consequent disputes.* The NEC is designed for international use and is being increasingly used in the UK and overseas.

GC/Works/1

The GC/Works/1 general conditions of contract were prepared for use on UK government contracts for building and civil engineering work. They are not used by all government departments or agencies. They are published in three versions

- *lump sum* with quantities
- lump sum without quantities
- single stage design and build version.

The latter is the most recent, and allows for any extent of design by the Contractor.

JCT Contracts

The Joint Contracts Tribunal is composed of organizations in the building industry. It has produced different sets of conditions of contract for use in building works, and these are widely used in the UK. The main JCT 'Standard Form of Building Contract' is published in different versions for private and local authority clients and for use with or without a bill of quantities.

The JCT has also published other conditions of contract for building projects, including

- with contractor's design
- intermediate form of building contract for works of simple content

---

* Sir Michael Latham in his review of construction in the UK recommended greater use of the NEC as it applies most of the principles he considered essential for modern contracts

- agreement for minor building works
- management contract and associated works contract and employer–works contractor agreement
- prime cost contract.

Institution of Chemical Engineers conditions
The Institution of Chemical Engineers has published model forms of conditions of contract for process plants in the UK. They are:

- lump sum (fixed prices) version (Red Book)
- reimbursable version (Green Book)

Model sub-contract conditions (Yellow Book) have also been published for use on reimbursable contracts.

The Green Book is for cost-reimbursable contracts for complete projects which include mechanical, electrical and process engineering, but it is being used for some high risk civil engineering contracts. It is also used as the basis for target-cost contracts for construction.

FIDIC International Conditions of Contract for works of civil engineering construction
These model conditions are recommended by FIDIC* for use where tenders for civil engineering work are invited on an international basis. They are based upon principles similar to the ICE conditions.

The FIDIC civil engineering conditions are in two parts. Part I comprises the General Conditions; Part II contains 'Conditions of Particular Application' which must be specially drafted to suit the particular contract. Explanatory material and example clauses are published to assist in preparing Part II. FIDIC have also published model conditions for international design-and-build and turnkey contracts. The FIDIC conditions are available in a number of different languages.

---

* Fédération International des Ingénieurs-Conseils, the international association of consulting engineers

Mechanical and electrical equipment
British, European and international model sets of conditions of
contract for the purchase of mechanical and electrical equipment
are listed in Appendix A.

Selection of contractors

Tendering
Some contracts are agreed by negotiation between a promoter and
one contractor. Much more usually, contractors are invited to tender
to undertake the work. The Promoter or his representative issues
the 'tender documents' that are the proposed basis for the contract.
For minor work these may consist of a sketch with a letter stating
when the work is required and simple conditions of payment. For
large or risky projects the tender documents usually consist of
detailed drawings, a specification and a bill of quantities or schedule
of payment, and the invitation to tender states that a specified set
of conditions of contract will apply.

Tendering may be open to all interested contractors, or limited to
some pre-selected list of contractors. Under open tendering, tenders
are invited by public advertisement. This system may appear to
provide the greatest competition, but it can have serious disadvantages. It can encourage contractors of inadequate experience
and resources to tender, and discourage firms who have the experience to foresee the risks and needs of the work. It also wastes the
resources used in preparing the unsuccessful tenders. The costs of
tendering are usually paid for by the tenderers, but must be reco-
vered in the prices paid to them on other contracts. Tendering for a
design-and-construct contract can be particularly costly, as it
requires the preparation and costing of a design by every contractor.
For some of these, the Promoters pay part of the tendering costs.

It is the policy of the UK government and most other experienced
promoters that tenders should be invited only from contractors
who have been pre-qualified for the type and risks of the project.
Contractors pre-qualify formally by submitting details of their rel-

evant experience, health and safety procedures, financial and managerial resources and other information such as performance on previous projects for the Promoter or others. Under selective tendering, a limited number of the pre-qualified contractors are invited to tender.*

In the public sector, it is also UK Government policy that a contract should be awarded to the tenderer who offers the best value for money, not necessarily the lowest tender price.[2] Private sector contracts are more often negotiated with trusted contractors who have previously performed in a satisfactory manner and who know what is expected.

EU Directives

For public works in the European Union tendering for many larger contracts must be carried out in accordance with procedures specified in EC Directives. Their purpose is to remove discrimination against contractors because of their nationality. Information on future contracts must be published in the official journal of the European Communities. Under these tendering rules, contractors invited to tender must be selected only from those who respond to the advertisement.

Qualified and alternative tenders

To permit comparison of tenders on a common basis, tenders are normally required to be submitted on the basis of the drawings, specification and general conditions issued in the invitation to tender, without reservations by a tenderer – these are called conforming tenders or 'without *qualification*'.

If qualified tenders are submitted and considered, the ideas offered and reservations stated by a tenderer should be evaluated systematically so that all the tenders can be compared fairly. A

---

* Sir Michael Latham's report '*Constructing the Team*' quotes recommendations that tenders should be invited from a maximum of six selected contractors, only four if for specialist work, and only three for design-and-construct contracts, plus perhaps two reserve names

tender in which a contractor offers an alternative design should be evaluated similarly. Usually alternative tenders are considered only if submitted together with a conforming tender.

Assessing tenders

Tenders should be assessed as a whole and checked systematically. For the purposes of comparing tenders, corrections should be made for arithmetical errors, particularly in the tenderers' prices, but the tenderers are not normally permitted to make corrections or other changes after the submission of tenders. Covering letters from tenderers should be examined, as these may include qualifications which are part of the offer and could be legally binding if not rejected.

A team of people should assess tenders, except for small projects, in order to bring together engineering, financial and other expertise. Using a team for this should also avoid the risk of accusations of personal bias in favour of one tenderer. The Project Manager or the consulting engineer who is to be the Engineer for the contract should be responsible for organizing this work and ensuring the delivery of a report to the Promoter recommending which tender (if any) to accept.

Selection of a management contractor

Selection of a management contractor can be as for a project manager or a contractor. Attention should be given to managerial capability and systems for project control.

Sureties – performance guarantees and bonds

A guarantee or bond in a contract establishes that if the Contractor fails to perform specified obligations, a third party (the guarantor) provides payment to the Promoter to compensate for that default.[3] Guarantees are often used

- as security that a tenderer will not withdraw his tender before a specified date

- as security that an advance payment by the Promoter to the contractor will be repaid if required by the contract
- as security against loss if a contractor fails to perform in accordance with the contract and the Promoter has to employ another to complete the Works.

Normally a Promoter will enter into a contract only with a contractor who has adequate financial as well as technical resources. He may not always have full information concerning a contractor's finances, his other commitments or previous record. Also, unforeseen factors may arise which affect the contractor's ability to perform the contract. Hence performance guarantees are a common requirement of many contracts.

The guarantor may be a bank, insurance company, the contractor's parent company or an independent financier. The Promoter should be satisfied that the proposed guarantor has the financial resources to satisfy the obligations in the guarantee. The liability of the guarantor is normally limited to a stated amount, often expressed as a percentage of the contract price.

'On-demand' bonds give the Promoter the right to claim payment from the guarantor without proof of default on the part of a contractor. Although these have been widely criticized, and have been the subject of litigation, they continue to be used in some countries.

Main contractors sometimes require similar guarantees from their sub-contractors. Promoters may also require collateral warranties from sub-contractors to provide a remedy for faults in a sub-contractor's work beyond the defects liability period of a main contractor.

References

1. EUROPEAN CONSTRUCTION INSTITUTE. *Client management and its role in the limitation of contentious claims*. 2nd edition, Loughborough, 1992
2. HMSO. *Setting new standards – A strategy for government procurement*. Cm 2840, 1995
3. EDWARDS L. S., LORD G. and MADGE P. *Civil engineering insurance and bonding*, 2nd edition, Thomas Telford, London, 1996

7

Planning and control of construction

Responsibilities

This chapter describes the main tasks of planning and controlling the construction stage of a project. Who is responsible for each task depends upon the contract arrangements. Fig. 11 indicates how these tasks are shared between the Promoter, Engineer* and Contractor in a traditional contract.

The tasks shown in Fig. 11 are more the responsibility of a contractor in design-and-construct contracts. They are completely a contractor's responsibility if he is promoting the project and undertaking its financing, design, construction and commissioning. In a contract between a promoter and a joint venture of contractors, the contractors are usually individually and jointly liable as if they were one contractor.

Initiation of construction

Starting date
In most contracts the Promoter or the Engineer notifies the Contractor in writing of the date for starting the Works. The Contractor is then responsible for proceeding with due diligence and completing the Works in accordance with his contract.

Under the UK CDM Regulations the Promoter has to appoint a

---

* or whatever title is used in a contract – see Chapter 4

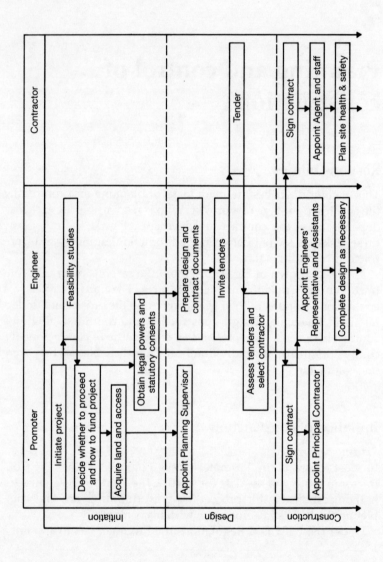

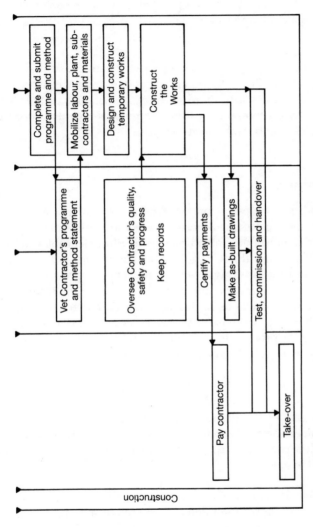

Fig. 11   Division of responsibility between Promoter, Engineer and Contractor

Principal Contractor and has the legal obligation not to permit construction to start until the Principal Contractor has prepared a satisfactory health and safety plan.

Statutory notifications
Statutory planning approval for a project is usually obtained by or on behalf of the Promoter. Contractors must meet legal requirements to notify new activity to statutory bodies such as the Health and Safety Executive, the local Environmental Health Officer and the Fire, Police and Ambulance Services.

The Engineer

In the traditional type of contract for the construction of a project the Promoter must name a person to be the Engineer. The person appointed then has duties and powers to administer the contract.

In the ICE 6 and the IChemE model conditions, the Engineer's decisions in administering a contract must be impartial as between Promoter and Contractor, and must be based on the terms of the contract. Any restrictions imposed by the Promoter on the Engineer's authority to exercise his powers stated in a contract must be notified to prospective contractors when inviting tenders, or negotiated between Promoter and Contractor during the contract as a change to its terms.

When supervising the work to see that the Contractor has executed his work in accordance with the contract, the Engineer is also the agent of the Promoter, thereby acting in a dual capacity. The Engineer normally reports to the Promoter monthly, including a review of the progress of the Works, major decisions and impending important events.

Programme and methods of construction
In the ICE conditions the Engineer is required to approve the programme of construction and consent to the methods which the Contractor proposes to use. These the Contractor must provide

under the contract if requested by the Engineer. This is called the Contract programme against which the progress of the Works is monitored.

Supervision

The Engineer usually appoints a representative to be on site (often called the *Resident Engineer*) to watch and supervise the construction and completion of the Works. The Engineer's Representative must look ahead and discuss future parts of the Works with the Contractor's manager in charge at the site to ensure that they are planned to achieve the approved programme. (*The Agent* is the traditional title in the UK for the Contractor's representative and manager in charge on site. The title Project Manager is now increasingly used by contractors, but is not used in this guide to avoid confusion with the Promoter's Project Manager).

Changes and variations

Figure 4 indicated that changes to design during construction can be expensive, because of the direct costs of repeating work and scrapping materials, and the indirect costs of disrupting economic working. In the traditional procedure in the UK the Promoter should not propose a change to the Contractor directly. It should be discussed with the Engineer and its benefits compared to cost, on the basis of its value in achieving the objectives of the project. If the Promoter then wishes to make the change, the Engineer may be able to instruct the Contractor to proceed with it, depending on the terms of the contract, or may have to negotiate it with the Contractor.

In the ICE conditions of contract the Engineer can instruct the Contractor to vary the Works if he thinks it necessary or desirable. Depending on the terms of the contract, variations may include instructions to add or omit work or change the contract programme. The Engineer should then instruct the Contractor by means of a formal Variation Order (VO). He should advise the Promoter on variations found necessary or desirable and inform the Promoter of their effects on the programme and the cost of the project.

VOs should specify the varied work in detail. The prices to be paid for new or additional work and other effects of a variation should be stated in a VO, if the Engineer has the power in the contract to do so, and if not, negotiated with the Contractor. VOs may be prepared by the Engineer's staff or representatives, but may be signed by them only if empowered to do so. In traditional admeasurement contracts, changes from the quantities of work stated in the bill of quantities which result from drawings issued by the Engineer during the contract are not 'variations'.

Changes proposed by the Contractor
A change proposed by the Contractor should be considered by the Engineer or his authorized representative and a recommendation whether or not to accept it made to the Promoter. If the proposal is accepted, the Engineer makes it a variation ordered under the contract. In some contracts the Contractor has a duty to propose variations which he considers may be necessary for the satisfactory completion of the project.

Changes negotiated between Promoter and Contractor
The parties to a contract can agree changes to it at any time, separately from powers that the Engineer may have to order variations. This is necessary for a change which the Engineer is not authorized by the contract to order as a variation.

Completion certificates
When, in his opinion, the Works have been substantially completed and passed the relevant tests, the Engineer is required to issue a certificate to that effect. The *defects correction period* then normally begins.

The Engineer may also at any time issue a completion certificate for a completed part of the Works. He must, if requested by the Contractor, issue a completion certificate for a substantial part of the Works if it is completed to the Engineer's satisfaction and is being used or occupied by the Promoter or anyone acting on his behalf or under his authority.

The Works must be handed over to the Promoter at the end of

the defects correction period in the condition required by the contract, and the Contractor must complete any outstanding work and also make good any defects during the defects correction period or immediately thereafter.

If any defects for which the Contractor is responsible are not corrected in this period, the Promoter is entitled to withhold from the balance of the *retention* money due to the Contractor the estimated cost of such work until the Contractor has completed it. Failing this the Promoter may arrange for it to be completed by others at the Contractor's expense.

Responsibilities of the contractor

The responsibilities of a construction contractor depend upon the terms of a contract and the relevant law. The following notes are written on the basis that one main contractor is 'the Contractor' in control of the site, but they also apply to each of several contractors if they are working on a project in parallel.

Implementation of the Works
The Contractor is responsible for constructing and maintaining the Works in accordance with the contract drawings, specification and other documents and also further information and instructions issued in accordance with the contract.

The Contractor should be as free as possible to plan and execute the Works in the way he wishes within the terms of his contract. So should sub-contractors. Any requirements for part of a project to be finished before the rest and all limits to the Contractor's freedom should therefore have been stated in the tender documents.

Health and safety
The Planning Supervisor appointed by the Promoter earlier in the project and the Principal Contractor's Agent should initiate meetings before construction starts to review the proposed methods of construction to identify hazards and minimize their dangers.

Representatives of all parties on a site should attend, and meet regularly to consider health and safety needs, plan preventative measures, arrange training and hear reports and recommendations on any accidents and near misses.

Successive contractors may need to be named as the Principal Contractor if possession of the site passes from one to another. On a multi-contract site the Promoter has to name one contractor as Principal Contractor to supervise health and safety over the whole site, as illustrated in Fig. 5.

Insurance
In most contracts the Contractor must insure the Works until they are handed over to the Promoter.[1] The cover is usually in the joint names of the Promoter and the Contractor.

Construction planning and control
Before the start of construction a scheme of work should be planned by the Contractor's senior staff who will be directly responsible for its execution. Decisions should be made on construction methods, site layout, temporary works, plant and the like, and requirements for labour, materials and transport.[2] The layout of temporary works areas, buildings, offices, accommodation, stores, workshops and temporary roads and railways needs attention, because the location of these features in relation to the Works can greatly affect the convenience and economy of future construction and administration.

Outline programmes prepared by the Contractor for tendering for a project are not likely to be based upon detailed study of the use of resources for the actual execution of the work. Detailed planning is normally needed at the start of construction in order to decide how to use labour, plant, materials, finance and subcontractors economically and safely.

As noted earlier, one of the first contractual duties of the Contractor is to submit a programme for the Engineer's approval. This programme should show the periods for all sections of the Works so that the Engineer can be satisfied that everything can be completed by the date specified in the contract. The Contractor is also required to submit a general description of his proposed method of work. If required by the Engineer, they must be amended by the Contractor and resubmitted at the earliest possible date.

The programme should show the Engineer when any further information, drawings or instructions will be required, and the dates when various sections of the Works will be completed and ready for use or for the installation of equipment by other contractors. All staff on site should review the programme and progress regularly to look ahead to check that the Works will be completed to the date specified in the contract.

Methods of programming
The most widely used forms of programme are bar charts and network diagrams. Bar charts (such as Fig. 2) can show programmes in a form that can be easily read and then used to compare progress with planned dates. Network diagrams show the sequence and interdependence of activities and indicate the effects of delays. Networks may be drawn as an arrow diagram or a precedence network. Either can be used to calculate the critical path of activities which determines the total time to complete all the work. A very simple example is shown in Fig. 12.

Many computer-based packages are available for displaying the network and the critical path of activities for large and complex projects. They can be used to analyse the use of resources, review progress and forecast the effects of changes in the timing of work or the use of resources. The choice of a computer package should be made after considering how far to integrate time and cost data for planning and control and also what is needed to feed back the resulting data into a database for planning and costing future contracts.

Note that 'programme' as used here means a diagram or table showing when work is to be done, as distinct from a computer software 'program'. 'Schedule' is an alternative name for a programme of work used particularly in the USA.

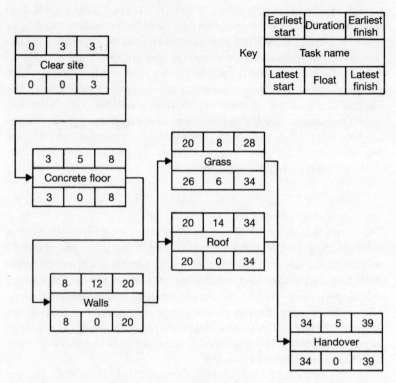

Fig. 12   Simple example of planning network

Detailed programmes

Every section of the project should have its detailed programme to ensure that the work is planned and methods and needs for materials are agreed in good time. Programmes should be limited in size to avoid confusion and to assist communication and understanding. Most computer-based planning packages can produce critical path diagrams and other results of data for 10 000 or so activities. To be able to understand the results it is usual to show no more than 50 activities in one programme, and to display the results in the form of a bar chart as illustrated earlier in Fig. 2. For all projects except small ones, the use of a hierarchy of programmes is therefore recommended, with one activity in a high level programme summarizing detail shown in a lower level programme.

Resource levelling
The Contractor will normally subject his draft programmes to a 'resource levelling' study to minimize the costs of peaks and gaps of the use of plant and labour. Computer-based systems are particularly useful for resource scheduling.

Labour planning
Contractors' estimates of the costs of work prepared for tendering are usually based on labour productivity for each type of work. These and the planned rate of work provide a basis for estimating the total labour. A chart of labour requirements can then be produced, showing by categories (skilled trades and labourers) the total numbers expected to be required at any particular time.

Economy in the use of plant and labour is achieved by planning to use it continuously and to maximum capacity. Planning and control should be in sufficient detail to see that expensive plant will not be idle for the want of adequate manpower or of transport of materials to or from the plant.

Plant planning
The construction plant that will be required and the periods during which it will be employed must be determined as early as possible, in conjunction with decisions on the detailed sequence of work and site layout. The time to obtain plant can be critical, depending on what plant is available from other contracts, what new plant should be purchased and what hired. For construction in many developing countries the need for supporting training and maintenance facilities has to be considered in deciding the choice of plant.

Programming the use of plant can be based on statistical data on the potential output for each machine and assessing the risks of interference and changes. As the work proceeds, data on actual output achieved with each major machine or set of machines should be analysed. The causes of poor output should be found, remedied, and a new forecast made of when the work will be completed. Records of output achieved should also form the basis of statistics for planning future work.

Materials planning
Materials typically account for 60% of the cost of construction. Their purchasing and use therefore need to be planned and controlled accurately, starting with detailed programmes which should enable buyers to draw up a schedule of materials required and ensure deliveries on time.

Sub-contracting planning
Sub-contractors should be appointed well in advance and their programmes obtained to show that they can perform their work properly in the time allowed.

Modifications to programme
Approval of a programme does not mean that it cannot be changed. A good programme is flexible enough to permit modifications to meet the more probable risks. Experience shows that a programme which allows for contingencies enables those in charge of the work to see what the effect of adverse events will be on subsequent work and adjust their plans accordingly. The working programme should therefore be updated regularly. Revisions will also be required if the Engineer varies the work, acceleration is required, or extra time is given by the Engineer for any reason. The Contractor should then submit a modified programme for the Engineer's approval.

Design of temporary works
The Contractor is usually responsible for the design of the temporary works he proposes to use, but these are of course subject to legal requirements for health and safety and specific regulations for independent checking of major falsework and other temporary structures.

In the ICE conditions of contract the Contractor is required to submit drawings and design calculations for temporary works to the Engineer. The Engineer should scrutinize these with care. This in no way relieves the Contractor of his responsibility for the design and construction of the temporary works, but it should reduce the risk of mistakes and helps the Engineer to discharge his responsibility to the Promoter to see that the work is done satisfactorily and safely.

Quality
Contracts usually specify in detail the quality of materials and workmanship required, and the tests required to prove compliance. When planning the construction work the Contractor should make sure that the proposed methods and plant can produce work to the quality specified. During the work the Contractor's senior management have to ensure that quality is achieved despite pressures to achieve progress in bad weather or other adverse conditions. Many Contractors have quality assurance systems which are designed to ensure this for all their work.

Setting out
The Contractor is usually responsible for setting out the Works. The Engineer's representative or assistants normally check the setting out in order to minimize the risk of errors and consequent delays to progress.

Reporting
The Contractor is normally required to report monthly on the progress and quality of construction and the supporting activities of ordering materials and designing temporary works. 'Returns' (data) of the number and classes of labour and plant employed on site by the Contractor and sub-contractors are usually also required, together with reports on any accidents and near misses.

Progressing
Assessments of progress should be based upon measuring or estimating the amounts of work completed, rates of work and trends,

so as to be able to predict when succeeding activities may start.[3]

Measurement of work done, rate of work and trends may be rough or detailed. Statements of elapsed time alone or man-hours or money expended are of little use, because they give no indication of what has been produced and when things will be completed. Data on progress should then be compared with predictions and the programmes to provide the basis for any action needed to achieve completion dates.

Cash flow
Every contractor and sub-contractor who is due to be paid according to progress of work needs to plan his expected cash flow when tendering, and then to monitor it during the work. Earlier expenditure or later payment than expected can incur greater costs for a contractor than his expected profit.

Cost monitoring system
On all but very small jobs every contractor and sub-contractor should have a site cost recording and monitoring system that provides

- accurate reports at regular intervals of the unit costs of all the principal work and overhead charges
- estimates of the likely final costs
- data in the form required for claiming progress payments
- cost data of the completed works to guide future estimating.

These data are needed whether the contract is fixed priced or costs are reimbursable. It is confidential to each contractor and sub-contractor unless required for payments under a reimbursable contract or part of a contract.

Labour and plant costs depend mainly on good planning and control on site. All these costs should therefore be recorded, reported and analysed, usually weekly, to show whether these resources were used productively, so that remedial action on problems can be taken before large wasteful costs have been incurred.

Monitoring of the costs of materials is needed to see whether materials are being used economically and to identify and remedy cases of waste and other losses. The accuracy of a costing system is very dependent on information from foremen, gangers, machine drivers and other operators on how man-hours have been used. The definitions of cost items should therefore be clear to them. The way that items of work are listed in a bill of quantities is not necessarily suitable for this purpose.

Each contractor and sub-contractor also needs an internal accounts system which provides reports and forecasts of expenditure, commitments, liabilities, cash flow and the expected final financial state of their contracts. Some or most of the cost monitoring and accounting required may be provided by a regional office or the company head office.

Payment to contractor and sub-contractors

Monthly statements for interim certificates

In most civil engineering contracts the Contractor is entitled to monthly payments for work completed. A statement (valuation) showing the amount due during the month in question is prepared and submitted by the Contractor to the Engineer (normally through the Engineer's Representative) in a form usually specified by the Engineer.

The statement should be checked and agreed by the Engineer's Representative and then forwarded to the Engineer who will, after verification, send to the Promoter a certificate showing the amount to be paid to the Contractor.

In a traditional admeasurement contract payment for most of the work done is based on the Contractor's tender rates for each item listed in the bill of quantities. Alternatively, the terms of the contract may be that payment is due when the Contractor completes defined stages of work (*milestones*). This basis of payment is becoming more common, as it is directly related to progress towards the Promoter's objectives. In cost-reimbursable contracts payment may depend upon agreement of the value of progress achieved, not simply the costs incurred. In most contracts the Promoter is obliged to pay the Contractor within a specified period (usually 28 days).

Final payment
The final balance of payment is usually due to the Contractor at the end of the defects correction period. By then the Engineer's Representative and the Engineer should expect a final account from the Contractor and be preparing to check it and issue the final certificate. If there remain outstanding claims from the Contractor, these should be settled under the procedure stated in the contract.

Claims and disputes

The procedure to be adopted in the settlement of disputes that may arise during a contract is usually established in the conditions of contract. In the ICE conditions, a matter does not contractually become a dispute until the Engineer has formally given his decision on it and one or other party to the contract (the Promoter or the Contractor) has formally objected to that decision. It may then be settled by negotiation between the parties (preferably conducted through the Engineer) or, if this fails, by reference to conciliation. If this is not successful, either party can refer it to arbitration.

The most common kind of dispute, where the Contractor has objected to a decision of the Engineer, is a claim from the Contractor for extra expenditure he has incurred, or for a loss he states that he has suffered because of circumstances which he considers he could not have been expected to allow for in his tender. If so, the Engineer must decide whether or not there is provision in the contract under which he can properly certify this expenditure for payment by the Promoter.

It is the duty of the Engineer's Representative and the Agent to record and if possible agree full details of the facts and cir-cumstances relevant to any matter that may be the subject of a claim. Where agreement on facts cannot be reached, separate records should be kept and the reasons for disagreement noted.

The assessment of a claim is a matter for the Engineer after discussion between him and the Contractor. Having ascertained all the relevant facts, the Engineer has to decide to what extent he considers a claim is justified under the terms of the contract. Having decided this, he should use factual data compiled from the records to price the claim and report his decision to the Promoter and the Contractor.

Dispute resolution

The use of conciliation to resolve a dispute is a term of the ICE 6 and Minor Works conditions of contract. Adjudication is the procedure in the NEC. The results of asking third parties or a law court to settle a dispute are uncertain, and the process can be very expensive. They should be used only after failure of the parties to agree by negotiation.

Following practice in the US, where litigation can add more than 10% to project costs, Alternative Dispute Resolution (ADR) procedures are being included in some contracts instead of immediate recourse to arbitration. ADR techniques are based upon the use of neutral experts. They require a change in contractual 'culture'. ADR methods vary, and can work successfully only if the parties in dispute are prepared to collaborate in solving a problem.

Communications and records

Communications between the Engineer or the Engineer's Representative and the Contractor should be in writing, or confirmed in writing.* Every document exchanged between the Contractor and the Promoter or Engineer on the execution of the Works – all letters, memos, drawings, sketches, photographs, data on disks, minutes of meetings, progress and other reports, monthly measure-

---

* *'Confirmation of verbal instruction'* forms for this purpose are available at the ICE bookshop

ments, claims and certificates – becomes part of the administration of the contract, unless agreed otherwise. All these and physical records of site conditions should be kept and filed to form the contract records.[4]

Documents whose importance or usefulness is not wholly clear at the time may become so later, for instance to help with such things as analysing the causes of accidents, failures or deterioration of completed work, the state of buried work, disagreements between the Engineer and the Contractor over payments or delays, and designing and pricing additional work. The records that should be kept include

- diaries – every engineer on a contract should keep a detailed diary of his own work, however insignificant the detail may seem at the time
- notes of oral instructions or agreements
- superseded drawings, which must be kept because something done when the drawing was current may seem very odd several years (and five revisions) later
- an up-to-date health and safety file, to guide all who will work on any future design, construction, maintenance or demolition of the project, and records of safety training and meetings
- as-built drawings – in practice the need to break into old work often reveals that the record drawings are out-of-date or incomplete.

Everybody may agree that good records should be handed over to the users of completed projects. Professional engineers have a duty to make sure that this happens.

References

1. EDWARDS L. S., LORD G. and MADGE P. *Civil engineering insurance and bonding*, 2nd edition, Thomas Telford, London, 1996

2. NEALE R. H. and NEALE D. E. *Construction planning,* Thomas Telford, London 1989
3. WEARNE S. H. (ed.) *Control of engineering projects*, Thomas Telford, London, 2nd edition, 1989
4. CLARKE R. H. *Site supervision*, Thomas Telford, London, 2nd edition, 1988

8

Construction management organization

Size and organization

The construction stage usually requires a large increase in the number of people employed on a project and the appointment of managers, engineers and supporting staff to site. This chapter describes typical organizations required to plan and control this stage of a project where there is one main contractor in the traditional civil engineering procedure in the UK and the work is supervised by the Engineer.* Most of the roles described would be in the Contractor's organization if responsible for design and construction.

The Engineer's organization

Project Engineer
Within the Engineer's organization an engineer is usually appointed the *Project Engineer* with these duties

- to plan and supervise design in detail, if not already complete, and coordinate the issue by the Engineer of any further drawings to the Contractor under whatever procedure is stated in the contract
- to direct redesign if there are varied requirements of the Promoter or changed site conditions, to estimate the

---

* or whatever title is used in a contract – see Chapter 4

effect which each of these variations will have on the programme and cost of the Works, and to advise the Engineer on issuing a Variation Order or whatever procedure is stated in the contract

- to advise the Engineer on the progress, trends and likely outcome of contracts
- to check the monthly measurement of work from the Engineer's Representative preparatory to the issue by the Engineer of interim payment certificates
- to administer the issue of the Engineer's certificates for payment to the Contractor
- to advise the Engineer on claims, disputes, completion and defects correction certificates
- to liaise with the Engineer's Representative on all the above.

Delegation of authority

Responsibility for most communications with the Agent is usually delegated by the Engineer to the Engineer's Representative. The extent to which the Engineer can delegate his powers in a contract is usually limited in the contract. Delegated decisions may also be subject to confirmation by the Engineer.

The Engineer should inform the Contractor in writing of the extent of delegation of his powers to the Engineer's Representative. The Engineer should also inform the Contractor of the names and positions of assistants to the Engineer and the Engineer's Representative he will have to deal with during the execution of the Works

The Engineer's Representative

The function of the Engineer's Representative is to watch and supervise construction. The principal duties of the Engineer's Representative are

- to check that the Contractor has organized his work to achieve the approved programme

- to examine the methods proposed by the Contractor for the execution of the Works, the primary object being to see that they should ensure safe and satisfactory construction
- to check that the Contractor and all others on site comply with health and safety requirements
- to assist the Contractor to interpret drawings and understand the specification, and refer questions to the Project Engineer
- to supervise the Works to check that they are being executed to correct line and level and that the materials and workmanship comply with the specification
- to assess and record the progress of the work in comparison with the programme
- to execute and/or supervise tests carried out on the site, and inspect materials and manufacture at source where this is not done by other representatives of the Promoter or Engineer
- to keep a diary constituting a detailed history of the work done and all events at the site and submit periodic progress reports to the Engineer
- to measure and agree as far as possible with the Contractor's staff the quantities of work executed, and to check *dayworks* and other accounts so that the interim and final payments due to the Contractor may be certified by the Engineer
- in the case of any work or event for which the Contractor may claim additional time or payment to agree and record the relevant facts before any question of principle has to be decided by the Engineer
- to record on drawings the actual level and nature of all foundations, the character of the strata encountered in excavation and details of any deviations from the drawings which may have been made during the execution of the Works
- to direct the production of as-built drawings

- to report to the Engineer on all the above and advise him on potential problems in good time for them to be avoided or their effects minimized.

The Engineer's site team

Except on small projects, the Engineer's Representative usually leads a team of assistant engineers and inspectors. Fig. 13 shows an example of an organization on site under the Engineer's Representative. Their actual numbers and organization depend upon the size of the project, the variety of the work and distance from head office or services. This example is typical of medium-size projects. In this example the team is divided into two main sections located at different parts of the site.

The role of the Inspectors is to supervise the Contractor's work, for instance the mixing and placing of concrete and any such work requiring constant supervision. The duties of Inspectors demand

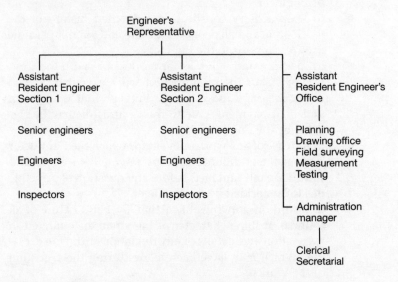

Fig. 13 On-site organization under the Engineer's Representative

practical experience, objectivity and tact in order to gain the respect of the foremen and skilled workmen employed by the Contractor.

Corresponding roles are needed in other contract arrangements, for instance where a project is designed and supervised by the Promoter's own staff.

Main contractor's organization

Figure 14 shows an example of a contractor's structure of company departments and the main responsibilities for managing contracts. In this example 'Project Director' is the title used for the Contractor's manager in charge of an exceptionally large project. There are four managers on site shown as responsible to the Project Director, but they are also accountable to the heads of functions who are located off-site in the company's head office or a regional office. The Contracts Managers are responsible for the company's normal range of projects (though not shown in Fig. 14, they and the Agents under them are also accountable to several heads of functions).

Figure 15 shows what might be a contractor's management structure on site for a medium-size project. Here some roles are specialist ones in a function or discipline, for instance planning. Others are in a section or area of the site. The nature of these duties varies significantly, depending on the variety of work, size and layout of a site and the terms of the contract. On major projects there tend to be more specialists; on smaller and traditional contracts there tends to be a wider range of responsibilities for each individual.

Contractors' Project Managers and Agents
Contractors' Agents are usually experienced engineers. Most of the financial risks of a construction contract are on site. The Agent is therefore usually given wide powers by his company to plan and control the work.

The Agent's main responsibilities are

- construction
- health and safety

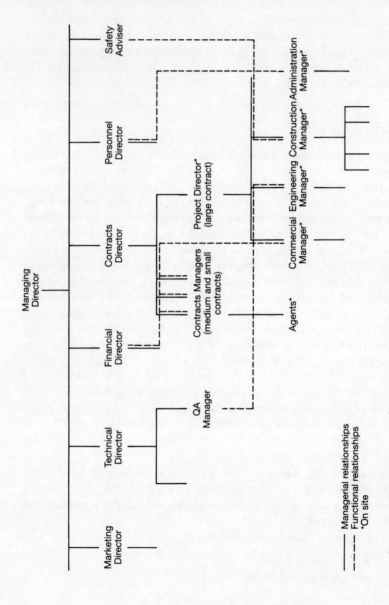

Fig. 14   Structure and responsibilities of a Contractor's company

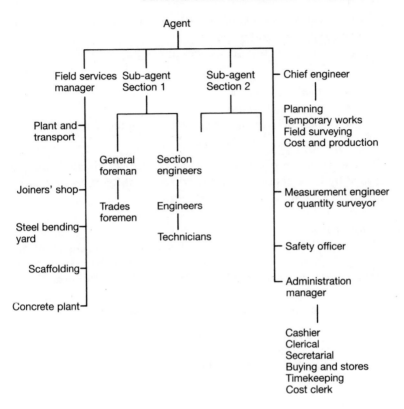

CONSTRUCTION MANAGEMENT ORGANIZATION

Fig. 15   On-site Contractor's management structure

- compliance with the contract
- the commercial success of the contract
- management of the Contractor's site staff
- liaison with the Engineer's Representative.

Sub-Agents

On larger sites areas of the work are usually the responsibility of Sub-Agents. Depending on the size of the particular project they will have varying numbers of staff, principally section engineers, assistant engineers and inspectors.

Sub-agents' main responsibilities are

- day-to-day site management
- health and safety
- industrial relations
- productivity and workmanship of the plant and labour
- management of transport.

The use of labour, plant and transport is usually planned and controlled by the General Foreman and sectional foremen, depending on the size, variety and spread of work.

General Foreman
The General Foreman is the link between the management and the foremen and gangers in direct charge of labour. His personal influence on the site organization can be a strong factor in achieving and maintaining efficiency.

The General Foreman's main responsibilities are

- allocation of labour to site operations
- supervising flows of materials and stores
- motivation of the labour force
- site communications between all foremen and gangers.

Chief Engineer or Engineering Manager (only on large projects)
The Chief Engineer is responsible for the planning and technical methods needed to ensure the quality and accuracy of the Works, through guidance to the Section Engineers and all the Contractor's staff. He has to assess the needs of drawings received from the Engineer and organize their issue on site. He is responsible for any design needed on site, especially of temporary works, drawing on head office engineering and research departments if appropriate.

Section Engineers
Section Engineers usually have experience of both design and con-

The construction plant that will be required must be determined as early as possible.

struction. They are ultimately responsible to the Agent but on larger sites are directed by the Chief Engineer on the quality and technical problems of the Works. They report progress and cost data to the Measurement Engineer. Each Section Engineer must liaise with his Foreman to plan the work of his section to be executed daily, weekly and monthly.

Measurement Engineer, Cost Engineer or Quantity Surveyor
On larger projects a separate Measurement Engineer, Cost Engineer or Quantity Surveyor and assistants may be needed to record the quantities of work done, check requests from sub-contractors for payment, and prepare the information for the Contractor to request payment as quickly as permitted under the contract.

Interim and final measurements and valuations and requests for payment for additional or varied work have to be substantiated and agreed with the Engineer's Representative. If agreement is not achieved, the Contractor may give notice of claims, in which case it is the duty of the Measurement Engineer to keep records for their later submission.

Normally the Measurement Engineer is also responsible for maintaining the records of correspondence and instructions on the main contract and sub-contracts, and advising the Agent on contract and sub-contract matters. On smaller projects these tasks are part of the job of Sub-Agents or Section Engineers.

Appointment of site staff
If the Works are to start well and proceed economically, much will depend on the early appointment of the Agent and his staff so that they can begin their planning and other preparations and establish selection, induction and training facilities before labour arrives in any numbers. There is a risk that these preparations will be inadequately planned if one person has to do several others' work during the initial period.

Selection of adequate and experienced staff and briefing them on the project, its priorities, risks and organization are particularly important if the site is remote – and especially if overseas.[1]

Industrial relations – communications and procedures

The *Working Rule Agreement* of the Civil Engineering Construction Conciliation Board for Great Britain negotiated between the employers' representatives and trades unions contains the terms of employment and grievance procedures for manual workers in the civil engineering industry.[2] Increasingly contractors and sub-contractors in the UK employ skilled and other labour as individual sub-contractors rather than as permanent employees, as this is more flexible for the employer and can have tax advantages to the employees, but the Working Rule Agreement provides a basis for their employment. Thorough knowledge of this Agreement and the amendments made to it from time to time is therefore needed for successful industrial relations.

Civil engineering in the UK has comparatively good industrial relations and care should be taken to maintain them. Good industrial relations on site are difficult to achieve and maintain unless management is seen to be consistent, fair and reasonable. To this end a good communication system should operate. Management should always be prepared to meet employees' representatives, resolve factual questions and explain policies.

Incentives

Contractors use incentive bonus schemes to try to achieve good productivity from manual employees. The basis of a good incentive scheme is that it should give a person of average ability the opportunity to earn more than his basic wage in return for increased production. Incentive schemes need to be seen to be fair. They require both technical and psychological skill to formulate and apply, otherwise discontent can quickly arise. Weekly measurement of production and the calculation of bonuses require promptness and accuracy.

Production targets and the bonus applicable to them should be clear and agreed between a contractor or sub-contractor and the representative of his employees, and thereafter altered only if circumstances justify changes.

Not all work can be made the subject of a bonus by the direct measurement of output. Employees who provide support services to those on bonus targets should therefore be given a financial interest in the work which attracts bonuses and so gain some benefit when bonuses are earned by their colleagues.

Site office administration
An Administration Manager (titles vary) on larger or remote sites is usually responsible for secretarial and other non-technical services to the site organization, the payment of wages, sickness records, minor purchasing, and the checking of the receipt and safeguarding of materials. On smaller and urban sites most or all of these services are provided by the Contractor's head or regional offices.

Sub-contractors' organizations

Sub-contractors' organizations are generally similar to those of main contractors, but smaller and more specialized to suit their scale and range of work.

References

1. LORAINE R. K. *Construction management in developing countries*, Thomas Telford, 1991
2. MARTIN A. S. and GROVER F. (eds), *Managing people*, Thomas Telford, London, 1988

9

Testing, commissioning and handover

Planning and organization

The period of testing, commissioning and handover of a project is the transition from its construction to its occupation by the end users. For the simplest projects this stage may consist of only a formal handover. For projects which include mechanical or other operating systems, controls and equipment, this stage is a separate set of testing and commissioning activities which overlap their supply and installation.

Planning

Testing, commissioning, handover and occupation requirements should be incorporated in planning activities from the earliest stages of a project, so that provision for them can be made in design and in deciding contractors' responsibilities. Together with the necessary budget provisions, this planning should be part of the project strategy. Suitable provisions can then be included in the relevant contracts with regard to responsibilities for testing, commissioning, partial and full handover, including commissioning by both the Promoter and the Contractor. This may need to allow for continued construction or fit-out work within a partially operational system or building.

Detailed planning of testing and commissioning activities will usually be necessary in the pre-commissioning stage, in parallel with construction.

Organization and resources
The Promoter may need to establish a commissioning and occupation team which includes his own representatives, others who are to be the eventual owners, users or occupiers, the engineering team and specialist commissioning personnel.

The resources needed should be identified and procured early in the project. An effective and active management structure should be established under the Project Manager, who should ensure that the responsibilities of each party are clearly defined and planned in order to ensure the smooth commissioning and handover of the project.

Health and safety
The testing, commissioning and operation of equipment and systems in a completed facility may include hazards with which construction managers and workers are not familiar. Other personnel from various organizations are likely to be working closely together, many not familiar with site hazards.

Health and safety responsibilities and the transfer of the control of hazards from construction to commissioning and operations staff should therefore be clearly defined and effective permit-to-work and other control procedures established. Detailed studies of risk and responsibilities, and training and induction in health and safety for all personnel during this stage are required.

Public liability
The safety of the public, as well as of occupiers and users should also be considered at all stages of commissioning.

Inspection and testing of work

Inspection and testing

During construction and on completion of parts of a project, inspection and testing are usually required in order to confirm compliance with the drawings and specification. Testing and inspection are generally required for static components of a project, while dynamic components such as machinery require testing and commissioning. The Project Manager or the Engineer, supported by his team, is usually responsible for inspections on and off-site and for testing of materials.

Test criteria and schedules

The performance tests and criteria to be applied to any aspect of work should be specified in the contract for that work as far as possible, so as to enable the identification of the state at which an acceptable quality or degree of completion has been achieved. Depending on the type of project, samples and mock-ups or factory inspections and acceptance tests may be required. If they are, the responsibility for their cost should be defined.

Schedules (lists) of the necessary inspections and tests should be agreed through collaboration between the Project Manager and all parties.

Commissioning

Commissioning roles

Commissioning should be the orderly process of testing, adjustment and bringing the operational units of the project into use. It is generally required where systems and equipment are to be brought into service following installation.

Commissioning may be carried out by the Promoter's or contractor's staff, by specialist personnel or by a mixed team. For complex industrial projects, a commissioning manager is usually appointed by the Promoter to plan the commissioning, preferably early in the project, establish budgets, lead a commissioning team and procure the other necessary resources. For simple projects, commissioning is usually undertaken by the Contractor, subject to the approval of procedures by the Promoter or Engineer. The commissioning of familiar or small process and industrial facilities is usually carried out by the operator with the advice and assistance of the suppliers of the equipment and systems.

Organization and management
Whatever the contractual relationships, the commissioning and occupation of a large project can involve a number of parties and risks. Active management, clear procedures and effective communication are thus essential. The management organization for pre-commissoning and commissioning should be agreed, and key personnel mobilized, during the construction stage, and organized so as to provide

- a simple structure to suit the specialized nature of the work, with clear responsibilities and without multiple layers of management.
- single point responsibility for all commissioning activities at all times, under the direction of a commissioning manager
- rigorous procedures in respect of health and safety.

Commissioning process
The commissioning process may need to allow for continuing construction alongside commissioning activities and, in many cases, occupation of premises or operation of completed systems. Thus, while construction proceeds on the basis of physical site areas and technical specializations, commissioning usually has to be undertaken on complete operating sub-systems and units.

Commissioning schedule
For all but simple static work the commissioning manager should draft a schedule listing all the items to be commissioned, their

interdependence and the standards of performance to be achieved. Commissioning is usually carried out progressively at the levels of sub-units, units, systems and the whole project. The schedule should provide for these activities being carried out sequentially, alongside continuing construction. Contingency plans should also be prepared, to allow alternative commissioning sequences if problems are encountered.

Commissioning plan

The commissioning plan should include a programme and identify all activities and the necessary resources and procedures. These should include the supply of power, water or other services for testing, materials, consumables, spares, labour and specialist expertise. The plan should also identify procedures for managing emergencies and for rectifying defects, both before and after handover.

Staffing and training

In order to ensure that the project is put into operation rapidly, safely and effectively, all the commissioning, operating and maintenance personnel should be appointed, trained and briefed before commissioning starts. This needs to be planned from project inception, so that the roles and activities of the commissioning and operating staff are integrated into a coherent team to maximize their effectiveness.

Completion and handover

Practical and sectional completion

In most construction contracts there is provision for partial or stage completion and handover of sections of the Works. Once the Contractor considers he has completed the scope of work within his responsibility and has fulfilled all the necessary obligations relating to a section or the whole of the Works, he may apply to the Engineer or Project Manager for a completion certificate.

Handover

For many projects, sectional completion may signal the handover of a physical unit from one contractor to another for further work, fit-out or equipment installation. Well defined procedures for such handovers should be established and agreed well in advance and should preferably form part of the relevant contractors' contracts.

Defects
Before acceptance of the Works and issue of a completion certificate, the Engineer or Project Manager should inspect the relevant works jointly with the Contractor and prepare a list of outstanding items of work or defects. Together with a programme for completion of the work, the schedule should be agreed with both the Contractor and the Promoter or a follow-on contractor, as appropriate.

Documentation
The commissioning and handover state is the point for finalizing the project documentation. There are three categories of documents to be handed over by designers, equipment suppliers and contractors

- records of the equipment and services as installed
- commissioning instructions, including safety rules
- operating and maintenance instructions, including safety rules.

Generally the first category will include design and performance specifications, test certificates, *snagging lists*, as-built drawings and warranties. In addition to these, documents for commissioning and operation will include permits, certificates of insurances, operation and maintenance manuals and handover certificates.

Acceptance, handover and certificates
Following achievement of the relevant performance test criteria, the Project Manager or Engineer is normally required to issue a certificate of completion, or partial completion as appropriate, to the Promoter. This is accompanied by the relevant outstanding work schedule and completion programme. For complex projects,

a handover certificate and detailed handover procedures may be used. Under many forms of contract, the issue of a completion certificate allows the release of part of the retention money held back by the Promoter from payments to the Contractor.

Warranties and defects liability
In the UK most of the Contractor's contractual responsibilities and general liabilities in respect of the works handed over pass to the Promoter (or to another contractor following on to do other work) upon issue of the completion or handover certificate. Thus, warranties for equipment, insurance liabilities and responsibility for operation, day-to-day cleaning, maintenance, health and safety may pass to the Promoter or follow-on contractor, as appropriate.

In many other countries, contractors and suppliers of equipment continue to have legal requirements to be insured against public liabilities.

Occupation

Planning
The occupation of a major development may, in itself, represent a significant project, requiring extensive planning and development, particularly for large purpose-built commercial facilities.[1] The occupation sequence may differ considerably from construction, in that it centres around employees themselves and involves both the style of management and the culture of the user's organization.

Consultation with representatives of users and employees can thus be important in project planning and design to identify their requirements and what facilities they require. Their involvement can also be important to ensure their commitment to the results and hence the ultimate success of the project.

Organization and control
To ensure a successful occupation sequence, it is not uncommon for the user to appoint a dedicated occupation or 'migration' project

team, headed by a project manager. Their responsibilities will include planning, programmes, budgets, methodologies, contingency plans, health, safety and control procedures and the identification of risks. Steering groups and representatives' groups may also be established to promote consultation and communication and assist with identification of requirements. The team will need to arrange support services and utilities and may also need to coordinate migration with continuing construction or fit-out work. They may also need to reconcile different interests of owners and occupiers.

Reference

1. CHARTERED INSTITUTE OF BUILDING. *Code of Practice for project management*, Ascot, 1993

10

Operation and maintenance

Operation and maintenance needs

A completed project becomes part of an operating system or public facility after handover. Some structures may then need only regular inspection and maintenance, but most become part of a system needing planning of its use as well as maintenance. Individual infrastructure facilities, for example a road or a bridge, require not only inspection and maintenance but also planning of their use in a highway system. Temporary closure for maintenance, refurbishment or replacement of components can affect traffic flows in a wide area. The operational tasks may also include the collection of tolls from users.

The annual cost of operating and maintaining static infrastructure facilities can be as low as 0.25% of the total construction cost. The cost to users if the facilities are not usable can be much greater. The annual cost of operating a process facility can be more than 10% of the total construction cost. Operations and maintenance therefore need to be planned during the design of projects, as discussed in Chapter 5.

Planning for operation and maintenance

Costs

The costs of operation and maintenance should be considered in the feasibility studies for a project, as mentioned in Chapter 3.

Differences in these costs between alternative schemes should influence the choice between them. Failure to consider these costs may lead to a choice of design that is uneconomic to maintain and operate or gives impaired performance during commissioning and operation.

Operational choices may also affect the capital cost. In a water treatment facility, for example, the choice of ozone instead of chlorine as a disinfectant will increase the construction cost but reduce the operating cost.

As-built drawings

As-built drawings should be prepared during construction, while detail can be seen and checked. For civil engineering work this is often undertaken by the Engineer's staff. For electrical, mechanical and process work it is frequently the Contractor or a sub-contractor who should prepare the drawings and submit them for approval by the Engineer.

At the end of construction, not later, a fully detailed set of drawings should be available to the operators, to represent the works as-built. They should include modifications made during the defects correction period. The drawings will be used to plan the use and control of the facility, maintenance and repair work, and as the basis of further design work for further development of the facility. The drawings handed over to the Promoter at the completion of the contract must therefore be a reliable representation of the actual Works at that time.

Manuals and operating documentation

The drawings and other documents must specify the operation and maintenance of hardware (equipment and structures), software, and the supporting needs of planning, stock control, billing and revenue collection, customer and employee relationships, training and career development.

Contracts for operation and maintenance

Increasingly the promoters of projects are employing other organizations to manage the operation and maintenance of completed facilities. These 'facilities management' contracts may be separate from those for design and construction, employing a following-on services contractor, or they may be part of a comprehensive contract for the project.

Process facilities, such as a water treatment plant comprising a reservoir, treatment facility and metered distribution network, require every component to have well defined contracts for their operation. The operational resources required, in the form of power, chemicals, manpower and the collection of revenue, can result in complex operation and maintenance programmes.

Figure 16 shows a procedure for defining the need for a maintenance agreement with a sub-contractor who has supplied and installed equipment for a project.

International funding agencies are frequently enthusiastic about including the responsibility for operations, maintenance and training (OMT) with the Contractor who supplied most or all of a project. The Contractor should then have a greater incentive to provide a facility which will be of high quality and operate effectively for a number of years, and ensure that skilled personnel will be available to train the Promoter's operating staff before transfer. The scope of OMT contracts can include many different combinations of operation, maintenance and training.

Operation only
Under these terms of contract the operator undertakes to operate and manage the facility, with any maintenance being the responsibility of the Promoter. A toll bridge, tunnel or leisure complex could be examples, where the operator would be primarily concerned with revenue collection but not be competent to carry out maintenance works.

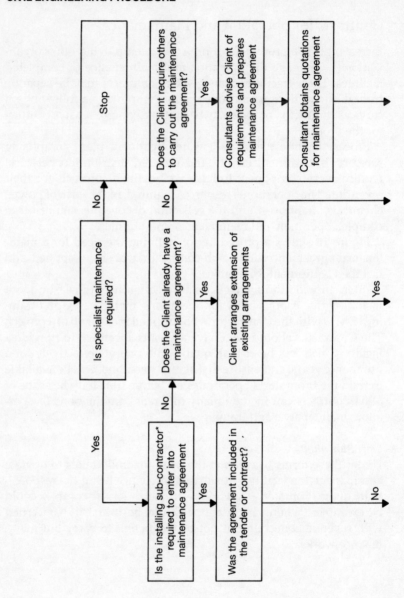

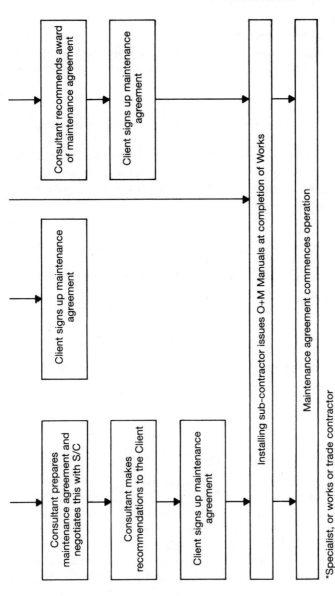

Fig. 16   Procedure for defining the need for a maintenance agreement

*Specialist, or works or trade contractor

Maintenance only

In this case the operator, usually the contractor, is responsible for only the maintenance of the facility. Examples may be the maintenance of particular equipment which may be required due to wear and tear over a fixed period.

Operation and maintenance only

Under these terms of contract, the operator undertakes to operate the facility and perform all routine and non-routine maintenance necessary to sustain the facility in full working order. In addition the operator is responsible for the provision of spares and consumables. An example may be the operation and maintenance of a water treatment facility.

Operation, maintenance and training (OMT)

Here the operator is obliged to train personnel, usually the Promoter's, to operate and maintain the facility until transfer. This form of contract may be considered for process or industrial plants which require high levels of operation and maintenance skills in order to ensure that revenue is generated during and after the contract.

Conditions of contract

OMT contracts thus differ significantly from construction contracts in their scope, duration, supervision, method of payment and the nature of risk, but most used by UK companies have been adapted from the ICE or FIDIC conditions of contract.

Training of operations personnel and managers

Training of operatives and managers may be part of an operation and maintenance contract. Training has two elements, the act of communicating knowledge and the act of receipt of knowledge. In many instances the latter is unfortunately disregarded as it can be difficult to quantify and assess in legal terms.

The first step in training must be to assess the existing skills of those who are to be trained. The training contract should clearly define how this should be achieved and against what criteria skills will be measured – in terms of quality as well as quantity. Trainees' effectiveness should be evaluated. The contract documentation must provide an appropriate mechanism for this.

The educational objectives of all training programmes should be specified in the contract. They should specify what the trainees will be expected to be competent to do, rather than what they should know. Job descriptions must be specific and unambiguous. Terms such as 'awareness' and 'understanding' should be avoided.

Training should not be considered to be an appendage to a contract designed to perform a dissimilar function. If training is to be included in the operation and maintenance contract, the human and material resources that will be allocated to training need to be defined. For the trainer, skills in interpersonal communication, sensitivity to different cultural norms and values, and a knowledge of training methods are as important as technical competence.

Overseas operating contracts

Current standards of operation and maintenance in many projects in both developed and developing countries are often far from ideal. The reasons vary, but common problems include the use of inappropriate design and technology, inadequate organizations, and the low priority and status Promoters give to maintenance work.

Care must be taken to assess the factors which may affect maintenance on overseas projects. Difficulties may sometimes arise due to the scarcity of trained managers, technicians and craftsmen, a shortage of foreign exchange or interruption of supplies, power, spares or raw materials. The division of responsibility for these factors needs to be clearly defined in a contract. Build-Own-Operate-Transfer (BOOT) and similar contracts are therefore favoured by some promoters to make the Contractor responsible for operating and maintaining systems and equipment that he chose and provided.

11

The future

Change

The fact that this is the fourth revision of this guide since its first publication in 1963 indicates how rapidly changes occur in the construction industry. Indeed the pace of change is accelerating in many countries. The previous revision of this book mentioned a variety of alternative methods of procurement and contract procedures being introduced into civil engineering. Most of these are now in use and have therefore been described in this revised edition.

The industry is likely to continue to change, and its future procedures determined by further developments in

- UK, EU and overseas markets
- public and private financing of projects
- technology
- European and world political changes
- environmental requirements
- conditions and methods of employment.

The professional engineer has to learn to deal with changes and to use them positively. Changes are opportunities for the advancement of organizations, the profession and the individual.

Changes in markets

Promoters' demands
Governments in the UK and other countries are introducing a

variety of procurement methods to try to attract private investment for the construction of new and improved transport systems and other infrastructure.

The privatization of former public services and nationalized industries means that they cease to be state agencies when promoting civil engineering projects. As commercial companies they have different demands, aspirations and pressures from their shareholders and their customers. This changes their trading patterns and requirements, and many have introduced alternative and innovative methods of procurement.

Public demands

The future needs for new projects cannot be predicted simply by extrapolating past growth. For instance, road construction and improvements already form a large market for civil engineering. The Department of Transport's figures indicate that road travel could increase by up to 140% by the year 2025, but the public are unlikely to agree that this trend should be answered only by constructing more roads. Society and industry may demand more coordinated road, tram and rail developments. Lack of public and private finance may limit new construction, and instead electronic tolling and other disincentives may be increasingly applied to motorways and other systems, particularly at busy times. Integrated information systems for journey planning and traffic management may become normal. Future transport policies may thus change the nature of new infrastructure projects. Civil engineers can expect substantial roles in planning, designing and constructing them, but with greater need for inter-disciplinary skills and cooperation in making complex decisions.

There is an increasing demand for new leisure and retirement facilities. The design and construction of these is growing in importance.

International markets

Maintenance of the existing infrastructures in the developed countries is a large volume of work. Future projects in these countries may be mainly extensions and improvements to existing facilities and smaller works for local social and economic reasons. Proceeding with many of these is limited chiefly by lack of private and government finance.

The former eastern European bloc countries need much investment in their energy, transport and services infrastructure. They hope for financial and other support from western Europe. Those countries have considerable construction expertise, and as their economies improve they will probably require technical and managerial expertise from the West.

The developing countries will continue to be a source of construction work, including some major projects financed by the international banks and aid agencies. Most countries in Africa, South East Asia, South America and much of the Pacific Rim are seeking more investment.

China is a massive potential market, but may want only technology transfer and some services from the West. India and Pakistan have well established construction industries, and will also mainly want technology transfer.

The industry

Competitiveness
Many factors can affect the successful execution of projects that satisfy customers. Of prime importance is delivering value for money. This demands innovative and excellent engineering, whether undertaken by Promoter, consultant or contractor, supported by safe and economic construction methods.

Designers and contractors are under increasing pressure from promoters and their financial supporters to reduce the costs of new projects and other work, in the UK and in other countries. To do so, everyone in the industry must continue to strive for improved efficiency in design and construction. The means of achieving this include

- best use of time, by individuals at all levels
- strategic use of information technology (IT), for predictions, modelling and managing design and construction as a manufacturing process
- greater use of systematic techniques for project selection, risk management and planning of the use of resources
- new materials such as fibre reinforcement, plastics and geotextiles
- further standardization and prefabrication of components for systems and structures
- mechanization of more labour-intensive site operations
- improved plant design and materials handling
- new technology such as satellite-aided surveying
- unified design, contract and construction data systems
- better use of value engineering and quality assurance systems
- much greater training and development at all levels, in technical, managerial and personal skills
- adoption of a continuous improvement culture and more systematic learning from project successes and failures.

and of course continuing research, development and innovation.

Health and safety
Inevitably sites are areas of risk. In the UK the law now forces all parties to anticipate and minimize these risks. The legal liability of promoters and designers to initiate health and safety programmes should greatly influence the health and safety attitudes and procedures of all parties from project initiation to operation.

Similar legislation applies in all the European Union and most industrialized countries. Standards and their enforcement may vary greatly from country to country, but professional obligations and possible new legal requirements to improve standards must be allowed for in planning future work.

Organizations

The efficiency and competitiveness of the civil engineering industry are increasingly important to individuals, companies and governments. To survive and succeed engineers and their employers should give more systematic attention to how people and their work are organized and controlled, within organizations and in contracts between them.

Organizations and their systems should be designed to be effective and motivating, not automatic or bureaucratic. Within an organization, the procedures and techniques which are appropriate vary from project to project, and are different for promoters, consultants and contractors. The professional engineer should be aware of the needs and be prepared to decide how the alternatives available should be applied to suit each project.

Implications for consulting engineers

The roles of consultants as specialists, experts, designers and project managers will continue and probably expand owing to more sophisticated design and construction methods. Increasing use of the New Engineering Contract system and greater reliance on contractors' quality assurance are changing the site role of the consulting engineer from supervisor to technical consultant.

Promoters increasingly want one organization to be responsible for all that is needed for a project. The effects of this are

- greater use of design-and-construct contracts, including design-build-finance-operate contracts for public projects in the UK and other countries
- increasingly multi-disciplinary consulting engineers' organizations in order to offer the range of skills needed to compete for employment on these projects
- mergers and acquisitions among consulting firms in order to provide the scale of resources needed for the larger projects and bear the increasing risks of competition.

Some promoters and contractors have acquired consulting engineer firms to be their in-house designers and project managers, while others have developed their own design and management teams.

These trends seem likely to continue. In addition, the UK and other governments may continue to sell their technical departments or establish them as consulting companies.

Implications for contractors

Pressures similar to those felt by consultants are leading to further reduction and rationalization amongst contractors. In response to the wider range of procurement arrangements demanded by promoters, many contractors are becoming project managers as well as constructors. They are having to develop multi-disciplinary teams capable of promoting, designing, managing and constructing projects.

Contractors from other European countries are established in the UK. This trend can be expected to continue, further increasing the competition for contracts. To achieve the economies of scale and the financial strength necessary to compete in UK and in overseas markets, UK-based contractors may have to work more in joint ventures or combine by merger or acquisition. This may be with other local contractors, or with those overseas to get better access to other markets. It may lead to the introduction of different management and procedural systems.

A much more complex industry is developing. Organizations and the responsibilities and authority of staff are becoming significantly more flexible and responsive to different priorities of projects. Management structures in many organizations are flatter to improve communications and reduce costs. These trends can greatly affect individuals and organizations, and there is no apparent reason why they should reverse.

Contracts

This guide describes the number and variety of general conditions of contract available in the UK and internationally. The problems that this produces are widely recognized. There is now considerable agreement that adversarial types of contract should no longer be used, resulting in a trend towards a single family of contracts for all construction.

There is likely to be a continuation of the trend away from resolving contract disputes by litigation and arbitration, towards using instead quicker and cheaper procedures such as mediation, conciliation and adjudication.

Partnering contracts are a development currently advocated by promoters and contractors, with these objectives

- reducing tendering times and costs
- anticipating and avoiding disputes
- securing contractors' best staff
- reducing duplication of management and supervision
- achieving quicker start to construction.

There are as yet no recognized model sets of conditions established in the UK or internationally for these contracts, but more of these contracts are probable and so consequently are developments in conditions of contract.

In the UK payment to contractors for the achievement of milestone activities on a critical path programme is tending to replace admeasurement based on bills of quantities. This releases engineers and quantity surveyors from the costly drudgery of detailed measurement and enables them to concentrate more on planning, anticipating problems, managing the use of resources and controlling variations.

Training
Investment in training is a characteristic of countries that achieve the fastest industrial growth. A much greater scale of training at all levels is required and needs to be sustained in the UK if it is to become more efficient and competitive in Europe and world markets.

To maintain an ability to provide economic and safe services, all engineers should keep up-to-date with the latest information through training and development and take the lead in using it. They should champion training and development for themselves and others, in order to continually improve professional knowledge, skills and attitudes, and they should show that this is being done.

Political and environmental requirements

Pollution
The 20th century has experienced an enormous growth in the world's population, and consequently in the demands for urbanization, transport, power and other polluting activities. Responding to these demands is now subject to ecological and other questions on the potential effects of atmospheric and other pollution, such as

- gas, chemical and smoke emissions – mainly due to road vehicles, domestic heating systems, power stations and factories
- noise and vibration – due to roads, aircraft, railways, factories and construction
- polluted water and contaminated land – due to industrial discharges, fertilizers, sewage and waste treatment and disposal
- visual – all construction and transportation, waste ground and neglected structures.

Many people in the developed world may agree to try to reduce pollution. The developing world and most former Eastern Bloc countries will have to continue these same polluting activities in order to maintain employment and services, let alone compete and develop. There is thus a major role for civil engineers in devising methods to satisfy these conflicting demands.

Sustainable development
As social and cost pressures grow to restrict the use of naturally available materials such as rock aggregates, engineers have to devise innovative economic methods for recycling old materials,

making better use of existing materials and using waste from previous mining and other activities.

Whole-life costing
The application of whole-life cost-benefit analysis to decisions in project design is changing the evaluation and selection of potential infrastructure projects. Engineers have to master and apply these techniques if they are to advise on projects and manage their design.

The engineering profession

Unification
The need for engineering teams and project managers with multi-disciplinary expertise, together with professional unification, may well result in the blurring of traditional distinctions between various professional skills.

Unification of the UK Engineering Institutions will create a single organization representing 270 000 Chartered Engineers, the largest professional body in the country by far. This should have considerable influence with government, and not only change the way that engineers relate within the profession but also improve the status and voice of the Chartered Engineer in society.

Establishment of a Construction Industry Council to represent all the UK professional organizations concerned with building, civil engineering and industrial construction has brought the Institution of Civil Engineers closer to its related Institutions and other bodies. This closer working within construction seems likely to continue, but become more complex with progressive integration of industrial and academic standards in Europe.

Continual improvement
Promoters increasingly expect that engineers will deliver their projects on time, at the right quality, lower cost and higher standards of health and safety. The public expect professional leadership to reduce the potential effects of new construction, changes to existing

structures and demolition work on the environment. Organizations and individuals expect more success in competing for work. These expectations can be achieved only by more rapid and successful innovations in design, construction, contracts and management.

The need for continuous professional development will never be greater. It is a matter of survival to develop and use new ideas and technology successfully. Understanding commercial, legal, environmental and organizational factors and the use of value engineering and quality assurance concepts is increasingly expected of most engineers. They should gain expertise in these before others take the lead in them. There is also a great need for engineers to acquire and develop a wider range of managerial expertise and personal skills.[1] A first degree is not an end in itself; for most engineers it is the start to acquiring an increasingly broader range of expertise and skills.

Reference

1. *Management development in the construction industry – Guidelines for the professional engineer*, Thomas Telford, London, 1992

Appendices

Canute: 'OK, OK, you win. Build your sea wall—just don't say I told you so.'

Appendix A

Model conditions of contract

Publications on UK practice and its problems are mostly written in terms of the use of various sets of 'model conditions of contract' for construction published by UK professional institutions, government departments and trade associations. These models are similar in their purpose of providing terms of contract to specify the responsibilities of the parties and the formal system of communications between them. They differ in the allocation of risks and in the detail of procedures and liabilities, and many clients add to them or have their own versions, but they are well known in construction in the UK and so probably influence clients' contract strategies. Listed here are the principal models which illustrate contract arrangements in practice. Many of these sets of terms are often called 'standards' in the UK, but 'models' is more accurate as they are not used uniformly.

ICE 6   *Conditions of Contract for Civil Engineering Works.* Institution of Civil Engineers, the Association of Consulting Engineers and the Federation of Civil Engineering Contractors, 6th edition, 1991 (ICE 6)

NEC   *The New Engineering Contract* family,* published by the Institution of Civil Engineers (ICE):

---

* The NEC family may be extended to include model conditions for product contracts, repairs and maintenance, and minor works

Engineering and Construction Contract, 1995
Engineering and Construction Subcontract, 1995
See also conditions of contract for professional services listed below.

FCEC *Form of Sub-contract* – for use with the ICE 6th edition. Federation of Civil Engineering Contractors, 1991

ICE Minor Works *Conditions of Contract for Minor Works.* Institution of Civil Engineers, 2nd edition, 1995

ICE Design and Construct *Conditions of Contract for Design and Construct.* Institution of Civil Engineers, 1992

ICE cond. for Ground Investigation *Conditions of Contract for Ground Investigation.* Institution of Civil Engineers, Association of Consulting Engineers and Federation of Civil Engineering Contractors, 1983

GC/Works/1 *GC/Works/1 Edition 3 General Conditions of Contract for Building and Civil Engineering.* Published by HM Stationery Office: Lump Sum with Quantities version 1989, revised 1990; Lump Sum without Quantities version 1991; Single Stage Design & Build version 1993. (Notes: (i) Previous editions were rather different; (ii) Government departments in the UK are not required to use these models).

JCT 80 *Standard Form of Building Contract.* Joint Contracts Tribunal for the Standard Form of Building Contract (JCT), 1980, plus amendments. Published in alternative versions for private and local authority clients, and with and without quantities or approximate quantities

JCT 81 *With Contractor's Design Contract.* JCT, 1981 edition, amended

JCT 84   *Intermediate Form of Building Contract.* JCT, 1984 (for smaller building projects)

JCT 87   *JCT Management Contract.* JCT, 1987

JCT Works   *Works Contract 3* (for use with the JCT Management Contract) JCT, 1987

IChemE Green Book   *Conditions of Contract for Complete Process Plants Suitable for Reimbursable Contracts.* Institution of Chemical Engineers, revised edition, 1992

IChemE Red Book*   *Conditions of Contract for Complete Process Plants Suitable for Lump Sum Contracts in the UK.* Institution of Chemical Engineers, revised edition, 1995. (Though indicated in the title of the above model, payment is rarely in a lump sum for large projects.)

IChemE Yellow Book†   *Conditions of Subcontract for Process Plants.* Institution of Chemical Engineers, 1992

MF/1   *General Conditions of Contract,* model form MF/1, for home or overseas contracts – with erection. The Institution of Mechanical Engineers, the Institution of Electrical Engineers and the Association of Consulting Engineers, 1988 (replacing previous model forms A and B3).

MF/2   *General Conditions of Contract,* model form MF/2, for home or overseas contracts – without erection. The Institution of Mechanical Engineers, the Institution of Electrical Engineers and the Association of Consulting Engineers, 1992

---

* FIDIC model conditions for international civil engineering contracts are also known as the 'Red Book'

† FIDIC model conditions for international mechanical and electrical contracts have for long also been known as 'the Yellow Book'

NEDO 1990 *Model Form of Contract for the Oil and Gas Industry.* The National Economic Development Office, 1990

There are other models produced by large promoters and trade associations and many sets of model terms for specialist work and sub-contracts.
Guidance notes are published with most of the above model conditions of contract.

Professional services contracts

ACE *Conditions of Engagement* – to form the basis for an agreement between client and consulting engineer for one or more stages of a project. Association of Consulting Engineers, revised 1995

APM *A Guide to Terms and Conditions for the Appointment of Client Project Managers* – with a schedule of duties & responsibilities for the construction industry. Association of Project Managers, 1994

NEC *Professional Services Contract.* Institution of Civil Engineers, 1994

NEC *Adjudicator's Contract.* Institution of Civil Engineers, 1994

International construction

FIDIC Red Book *Conditions of Contract (International) for Works of Civil Engineering Construction.* Fédération International des Ingénieurs-Conseils (FIDIC), 4th edition, 1987, revised 1988. (This model has many similarities to the ICE 5 model conditions of contract, but also some differences.)

FIDIC Yellow Book *Conditions of Contract (International) for Electrical and Mechanical Works.* Fédération International des Ingénieurs-Conseils (FIDIC), 3rd edition, 1988. (This model has many similarities to the IMechE/

IEE/ACE MF/1 model conditions of contract, but also some differences.)

**FIDIC
Orange
Book**
Conditions of Contract (International) for Design-Build and Turnkey. Fédération International des Ingénieurs-Conseils (FIDIC), 1995

ECE
General Conditions for the Supply of Plant and Machinery for Export. United Nations Economic Commission for Europe (ECE) (alternatives for supply with and without erection)

EDF
General Regulations for Works, Supply and Services Contracts Financed by the European Development Fund. European Development Fund, 1990
General Conditions for Works Contracts. European Development Fund, 1990
General Conditions for Supply Contracts. European Development Fund, 1990
General Conditions for Service Contracts. European Development Fund, 1990
Procedural Rules for Conciliation and Arbitration of Contracts. European Development Fund, 1990

The EDF publications listed above are published in the Official Journal of the European Communities No. L382 dated 31 December 1990.

Definitions used in model conditions of contract

Table A.1 lists some of the equivalent definitions used in the more common model sets of conditions of contract for larger projects. These are approximate equivalents, dependent on their definition in the particular conditions of contract.

The numbers indicate the clauses where these words are defined or first used in the set of conditions. For the NEC the numbers refer to the core clauses except where option A is indicated.

Table A.1   Definitions used in model conditions of contract

ICE 6 and FIDIC (Civil engineering)	NEC	GC/Works/1	JCT	IChemE (Red and Green Books)	MF/1 and MF/2
Employer 1	Employer 11.2	Authority 1	Employer 1.3	Purchaser 1	Purchaser 1.1
Contractor 1	Contractor 11.2	Contractor 1	Contractor 1.3	Contractor 1.1	Contractor 11.1
Engineer 1	Project Manager 11.2 Adjudicator 11.2	PM 1 Adjudicator 59	Architect or Contract Administrator 1.3	Project Manager 1	Engineer 1.1
Engineer's Representative 1	Supervisor 11.2	Clerk of Works or Resident Engineer 4	Clerk of Works 1.4	Project Manager's Representative 1	Engineer's Representative 1.1
Contractor's agent 15		Agent 5	Person-in-charge 10	Site Manager 1	Contractor's Representative 17.1
Tender total 1	Prices A 11.2	Contract sum 1	Contract sum 14.7	Contract price 1	Contract price 1.1
Contractor's equipment 1	Equipment 11.2	Contractor's plant 15	Contractor's plant 14.5	Contractor's equipment 1	Contractor's equipment 1.1

Table A.1   Definitions used in model conditions of contract (continued)

ICE 6 and FIDIC (Civil engineering)	NEC	GC/Works/1	JCT	IChemE (Red and Green Books)	MF/1 and MF/2
Interim certificate 60	Payment certificate 51.1	Advances on account 48	Interim certificate 1.3	Certificate for an instalment 39	Interim certificate 39.1
Final certificate	Payment certificate after completion of the whole of the works	Final certificate for payment	Final certificate	Final certificate	Final certificate of payment
60	50.1	50	1.3	1	39.9
Certificate of substantial completion 1	Certificate of taking over 35.4	Certificate 39	Certificate of practical completion 17.1	Taking-over certificate 34	Taking-over certificate 29.2
Defects correction period 1	Defects correction period 40.5	Maintenance period 21	Defects liability period 17.2	Defects liability period 36	Defects liability period 1.1
Defects correction certificate 1	Defects certificate 11.2	Certificate 39	Certification of completion of making good defects 17.4	Final certificate 1	Final certificate of payment 39.9

Appendix B

Bibliography

This list should be used selectively, depending upon individual needs and future interests. No one is expected to read all of these publications. On some subjects there are alternatives. These and any new publications since this list was compiled may be available in the Institution's or other libraries and are usually on sale at the Institution's bookshop.

The Promotion of projects
CORRIE R. K. (ed.), *Project evaluation*, Thomas Telford, London, 1990
SCANLON J. B. H. *Marketing of engineering services*, Thomas Telford, London, 1988
SHAUGHNESSY H. (ed.), *Project finance Europe – New public and private sources*, John Wiley & Sons, Chichester, 1994
MERNA A. and SMITH N. J. (eds), *Projects procured by private financed concession contracts*, Oriel, Manchester, 1994
FORDE M. C. (ed.), *Polluted and marginal land*, Thomas Telford, London, 1992

Risk management
EDWARDS L. J. *Practical risk management in the construction industry*, Thomas Telford, London, 1995
SKIPP B. O. (ed.), *Risk and reliability in ground engineering*, Thomas Telford, London, 1993

THOMPSON P. A. and PERRY J. G. (eds), *Engineering construction risks – a guide to project risk analysis and risk management*, Thomas Telford, London, 1992

HM Treasury Central Unit on Purchasing, *Risk*, HMSO, London

Responsibilities for a project

COX P. A. (ed.), *Civil engineering project procedure in the EC*, Thomas Telford, London, 1991

STALLWORTHY E. A. and KHARBANDA O. P. *Guide to project implementation*, Institution of Chemical Engineers, Rugby, 1986

HURU M. *The UK construction industry: A continental view*, Construction Industry Research & Information Association, London, 1992

INSTITUTION OF CIVIL ENGINEERS, *Functions of the Engineer under the ICE conditions of contract*, Guidance note 2A, Conditions of Contract Standing Joint Committee, Institution of Civil Engineers, London, 1977

Design

Design guides – series being published by Thomas Telford, London

RUTTER P. A. and MARTIN A. S. *Management of design offices*, Thomas Telford, London, 1990

Organization

CONSTRUCTION INDUSTRY COUNCIL, *The procurement of professional services*, London, 1993

MARTIN A. S. and GROVER F. (eds), *Managing people*, Thomas Telford, London, 1988

WEARNE S. H. *Principles of engineering organization*, Thomas Telford, London, 2nd edition, 1993

Project management

LOCK D. *Project management*, Gower Publishing Co, London, 6th edition, 1996

RUSKIN A. M. and ESTES W. E. *What every engineer should know about project management*, Marcel Dekker, New York, 2nd edition, 1994

SMITH N. J. (ed.) *Engineering project management*, Blackwell, 1995

WEARNE S. H. (ed.) *Control of engineering projects*, Thomas Telford, London, 2nd edition, 1989

CHARTERED INSTITUTE OF BUILDING, *Code of practice for project management*, Ascot, 1993

BADEN HELLARD R. *Total quality in construction projects*, Thomas Telford, London, 1993

Health, Safety and Welfare

Health & Safety Commission guidance notes on the *Construction (Design and management) Regulations* and other legislation

INSTITUTION OF CIVIL ENGINEERS, *Managing health and safety in civil engineering*, Thomas Telford, London, 1995

Contract Strategy

CURTIS B. *et al. Roles, responsibilities and risks in management contracting*, Publication SP81, Construction Industry Research & Information Association, London, 1991

EUROPEAN CONSTRUCTION INSTITUTE, *Client management and its role in the limitation of contentious claims*, Loughborough, June 1991, revised April 1992

EUROPEAN CONSTRUCTION INSTITUTE, *Construction contract arrangements in EU countries*, Loughborough, 1993

HODGSON G. J. Design and build – Effects of contractor design on highway schemes, *Civil engineering. Proc. Instn Civ. Engrs*, 1995, **108**, pp. 64–76

SIR MICHAEL LATHAM, *Constructing the team*, report of the joint government industry review of procurement and contractual arrangements in the United Kingdom construction industry, HMSO, London, 1994

NATIONAL ECONOMIC DEVELOPMENT OFFICE, *Faster building for industry*, London, 1983

National Economic Development Office, *Partnering: Contracting without conflict*, London, 1991

Perry J. G., Thompson P. A. and Wright M. *Target and cost-reimbursable construction contracts*, report 85, Construction Industry Research & Information Association, London, 1982

Wright D. *An engineer's guide to the model forms of conditions of contract for process plant*, Institution of Chemical Engineers, Rugby, 1994

Contract Management

Atkinson, A. V. *Civil engineering contract administration*, Stanley Thornes Publications, 2nd edition, 1992

Boyce T. *Successful contract administration*, Hawksmere, London, 1992

The management of project information, PROP1340IT series papers, Construction Industry Research & Information Association, London, 1995

Edwards L. J., Lord G. and Madge P. *Civil engineering insurance and bonding*, Thomas Telford, London, 2nd edition, 1996

Haslam J. R. *Specifications*, Spon, London, 1988

Horgan M. O'C. and Roulston F. R. *The foundations of engineering contracts*, Spon, London, 1989

Guidance on the preparation, submission and consideration of tenders for civil engineering contracts, Institution of Civil Engineers, London, 1983

Manson K. *Law for civil engineers*, Longman, London, 1993

Uff J. F. *Construction law*, Sweet & Maxwell, London, 6th edition, 1994

Wearne S. H. (ed.), *Civil engineering contracts*, Thomas Telford, London, 1989

Contract Supervision

Clarke R. H. *Site supervision*, Thomas Telford, London, 2nd edition, 1988

Elsby W. L. *The engineer and construction control*, Thomas Telford, London, 1981

Construction Management

HARRIS F. and McCAFFER R. *Modern construction management*, Blackwell Scientific Publications, Oxford, 4th edition, 1995

NEALE R. H. and NEALE D. E. *Construction planning*, Thomas Telford, London, 1989

SHAUGHNESSY H. (ed.) *Collaboration management – New project and partnering techniques*, John Wiley & Sons, New York, 1994

Works construction guides published by Thomas Telford, London

Cost Control

BARNES N. M. L. (ed.) *Financial control*, Thomas Telford, London, 1990

SMITH N. J. (ed.), *Project cost estimating*, Thomas Telford, London, 1995

McCAFFER R. C. and BALDWIN A. N. *Estimating and tendering for civil engineering works*, Blackwell Scientific Publications, Oxford, 2nd edition, 1991

Engineers and auditors, Institution of Civil Engineers and Chartered Institute of Public Finance and Accountancy, London, 1992

Overseas Projects

LORAINE R. K. *Construction management in developing countries*, Thomas Telford, London, 1991

Overseas projects – Crucial problems, Conference, Institution of Civil Engineers, London, 1988

Management of international construction projects, Conference, Institution of Civil Engineers, London, 1985

Professional Duties

Reviewing the work of another engineer, Institution of Civil Engineers, London, 1991

Guidance memorandum for expert witnesses and their clients, Institution of Civil Engineers, London, 1988

Management development in the construction industry – Guidelines for the professional engineer, Thomas Telford, London, 1992

McDONALD STEELS H. *Effective training for civil engineers*, Thomas Telford, London, 1994

CLAYTON R. (ed.), *Guide to consultancy*, Institution of Chemical Engineers, Rugby, 2nd edition, 1995

MITCHELL J. *How to write reports*, Fontana, London, 1974

SCOTT W. *Communication for professional engineers*, Thomas Telford, London, 1984

Appendix C

Useful addresses

Institution of Civil Engineers
Great George Street
London
SW1P 3AA

Association of Consulting
Engineers
Alliance House
12 Caxton Street
London
SW1H 0QL

Association of Project Managers
85 Oxford Road
High Wycombe
Bucks
HP11 2DX

Building Employers
Confederation
82 New Cavendish Street
London
W1M 8AD

Chartered Institute of Building
Englemere
Kings Ride
Ascot
Berks
SL5 8BJ

Construction Industry Council
26 Store Street
London
WC1E 7BT

Construction Industry Research
& Information Association
6 Storey's Gate
Westminster
London
SW1P 3AU

European Construction Institute
Sir Arnold Hall Building
University of Technology
Loughborough
Leics
LE11 3TU

Federation of Civil Engineering
Contractors
Cowdray House
6 Portugal Street
London
WC2A 2HH

Institution of Chemical
Engineers
Engineering Department
London
Davis Building
165 Railway Terrace
Rugby
CV21 3HQ

Centre for Dispute Resolution
100 Fetter Lane
London
EC4A 1DD

Chartered Institute of
Arbitrators
24 Angel Gate
City Road
London
EC1V 2RS

Chartered Institute of
Purchasing & Supply
Easton House
Easton on the Hill
Stamford
Lincs
PE6 3NZ

Institution of Electrical
Engineers
Savoy Place
London
WC2R 0BL

RIBA Publications
39 Moreland Street
London
EC1V 8BB

Thomas Telford Services Ltd
1 Heron Quay
London
E14 4JD

Appendix D

Glossary

Admeasurement – apportioning of quantities or costs. See also
Remeasurement, Valuation, Bills of quantities

Adjudicator – in the New Engineering Contract system the person
appointed to give a decision on a dispute between the parties to
the contract

Agent – in civil engineering in the UK 'Agent' is traditionally the
title of a contractor's representative on a site. Many now have the
title 'Project Manager' (see below)

Bid – an offer to enter into a contract; such an offer by a contractor
to construct a project. Also called a 'Tender'

Bills of quantities (BoQ) – a list of the items and quantities of delivered
work to be done for the Promoter under a contract, for instance
a quantity of concrete placed to a specified quality. An equivalent
in some industrial contracts is called a schedule of measured
work. See also *Schedule of rates, Unit rate*. In the traditional
arrangements a BoQ is normally issued with an invitation to
contractors to tender for a project, with a specification and the
tender drawings, and the contractors insert their 'rates' (prices
per unit quantity) for each item. The rates in the contractors'
tenders can then be compared item by item. In what are called
'admeasurement' contracts, payment to a contractor is based on
these rates × the final quantities of work done

CDM (formerly CONDAM) – the Construction (Design &
Management) Regulations, 1994. UK legal requirements

implementing European Commission Directive 92/57/EEC

Client – see *Promoter*

Conditions of contract – the 'conditions' of a contract means all the important terms of that contract. In civil engineering the words 'conditions of contract' are used to mean sets of contract terms on general matters likely to be required in all contracts for a class of projects. They define the words used, the responsibilities of the parties, procedures, liabilities for damage, injuries, mistakes or failures of contractor or sub-contractors, delays, changes in legislation such as taxation, frustration of contract and termination. They are designed to be used with a specification, drawings, schedules and other documents which state the particular terms of a contract. They are also called 'general conditions of contract', 'model forms' or 'standard forms'. See also *Terms of a contract*

Construction management – is used with a special meaning in the USA and the UK to mean the employment by a Promoter of a construction contractor to help plan, define and coordinate design and construction, and supervise construction by 'Works contractors' (see below)

Consultants – professional advisers on studies, projects, design, management, techniques and technical or other problems

Consulting engineers – consultants who also design projects and supervise construction, usually in firms of partners and supporting staff

Contract – an agreement enforceable at law

Contract price adjustment/fluctuation – a term in some contracts for adjustment of the contract price for the effects of inflation on the costs of labour and materials. Data produced from public records of the increased cost of commodities, labour, fuel and so on (such as the Baxter and Osborne indices in the UK) can be used to ascertain the amount due in each month's payment. See also *Escalation*

Contractor – in general a supplier of services; in civil engineering usually a company which undertakes the construction of part or all a project, and for some projects also undertakes design or other services

The Contractor – a party to a contract, as distinct from any contractor. The word 'the' is important in English practice as identifying the particular contractor who has entered into a contract for a project. The Contractor may be a joint venture of two or more companies

Cost plus – see *Reimbursable contract*

Dayworks – see *Schedule of rates*

Defects correction period or *Defects liability period* – see *Retention*

Direct labour – a Promoter's own employees employed on construction, sometimes under the internal equivalent of a contract, otherwise as a service department

Domestic sub-contractor – a sub-contractor selected and employed by a main contractor, i.e. not nominated by the client

The Employer – see *Promoter*

The Engineer – a person named in a contract to be responsible for administering that contract, particularly in contracts for construction or for the supply and installation of equipment

The Engineer's Representative – in ICE contracts the formal title for the Engineer's representative on site, often called 'the Resident Engineer' – see also *Supervisor*

Equipment – machines, services and other systems. 'Contractor's equipment' is defined in some civil engineering contracts to mean things used by the contractor to construct the works but not materials or other things forming part of the permanent works – see also *Plant*

Escalation – increases (or decreases) in the costs of labour or materials due to inflation (or recession and deflation). See also *Contract price adjustment*

Feasibility studies – investigation of possible designs and estimating their costs to provide the basis for deciding whether to proceed with a proposed project

Firm price – varies in its meaning, but is often used to indicate that a tendered price is offered only for a stated period and is not a commitment if it is not accepted within that period – see below

Fixed price – usually means that a tender price will not be subject to escalation, but it may mean that there is no variations term

NB Like some other words used in contract management, 'fixed price' has no fixed meaning and 'firm price' no firm meaning. What matters in each contract is whether its terms of payment include provisions for changes to the contract price, and what the governing law permits

General contractor – a contractor who undertakes the whole of the construction of a project, but usually in turn sub-letting parts of his work to specialist or trades contractors and others as sub-contractors

Liabilities – legal obligations

Lump sum payment – used in engineering and construction to mean that a contractor is paid on completing a major stage of work, for instance on handing over a section of a project. Strictly it means payment in a single lump. In practice 'lump sum' is used to mean that the amount to be paid is fixed, based on the contractor's tender price but perhaps subject to contract price adjustment

Main contractor – similar to a general contractor: a contractor who undertakes the construction of all or nearly all a project, but usually in turn sub-letting parts of his work to specialist or trades contractors and others as sub-contractors

Maintenance period – older name for 'defects liability period' – see *Retention*

Management contractor – a contractor employed by the Promoter to help plan, define and coordinate design and construction and to direct and supervise construction by other contractors – see *Works contractor*

Managing contractor – a main contractor, as distinct from a 'management' contractor

Measurement – calculation of quantities of work for a BoQ or for payment – see also *Remeasurement*

Milestone and *planned progress payment* – payment to a contractor in a series of lump sums each paid upon his achieving a 'milestone' – meaning a defined stage of progress. Use of the word milestone usually means that payment is based upon progress in completing what the Promoter wants. Payment based upon achieving defined percentages of a contractor's programme of activities is also known as a 'planned payment' scheme

Model conditions of contract – see *Conditions of contract*

Nominated sub-contractor – a sub-contractor usually for specialist work who is chosen by the Promoter or the Engineer rather than by the main contractor but is then employed by the main contractor

Owner – see *Promoter*

Partnering – collaborative management of a contract by promoter and contractor to share risks and rewards

Permanent works – the works to be constructed and handed over to the Promoter

Planned payment – see *Milestone*

Planning Supervisor – the person a Promoter is required under the CDM Regulations (see above) to appoint to ensure that a health and safety plan is prepared, adjusted as necessary and applied to the design and construction of a project

Plant – traditionally 'Contractor's plant' is defined in civil engineering contracts to mean things used by the contractor to construct the works but not materials or other things forming part of the permanent works – see also *Equipment*

Pre-qualification – the process of inviting the consultants, contractors or sub-contractors who are interested in tendering for work first to submit information on their relevant experience, performance, capacity, resources, systems and procedures, and from this information assessing which are qualified to be invited to tender – see also *Qualification*

Principal – see *Promoter*

Principal contractor – the contractor appointed under the CDM Regulations (see above) to be responsible for ensuring compliance with the health and safety plan by all contractors and individuals on construction – usually the main or largest contractor on a site

Project – any new structure, system or facility, or the alteration, renewal, replacement, substantial maintenance or removal of an existing one

Project Engineer – the title often used in promoters' and consultants'

organizations for the role of an engineer responsible for leading and coordinating design and other work for a project. In some cases this title is used where the role is actually the greater one of Project Manager as described below

Project management – sometimes used with the special meaning that the Promoter employs an independent professional project manager to help plan, define and coordinate the work of consultants and contractors to design and construct a project

Project Manager – the title increasingly used in promoters', contractors' and consultants' organizations for the role of manager of the development and implementation of a project. The role may have other titles, such as 'Project Director' for a large project. For smaller projects the role is not necessarily a separate job. In this guide the title Project Manager is used only for this role in a Promoter's organization

The Promoter – the 'client' for a project, the individual or organization that initiates a project and obtains the funds for it. In some contracts the Promoter is called 'the Employer', in others 'the Owner', 'the Purchaser', 'the Principal'

Provisional sums – amounts included in a BoQ for work which is not defined before inviting tenders. The contractor is paid an amount based upon the actual work ordered by the Engineer

Punch list – a list of defects to be corrected by a contractor or subcontractor

Purchaser – see *Promoter*

Qualification – commonly used to mean an accomplishment or attribute which is recognized as making a person or an organization fit to undertake a specified role or function – see also *Pre-qualification*
– also used to mean that a tender includes reservations or statements made to limit liabilities if that tenderer is given the contract

Qualified tender – a tender which includes reservations or statements made to limit liabilities if that tenderer is given the contract – see above

Rate – price per unit quantity of an item of work. Not used to mean the speed of work

Reimbursable contract – a contract under which a promoter pays ('reimburses') all a contractor's actual costs of all his employees on the contract ('payroll burden') and of materials, equipment and payments to sub-contractors, plus usually a fixed sum or percentage for management, financing, overheads and profit. This is often called a 'cost plus' contract

Remeasurement – calculation of the actual quantities of work ordered on the contractor in order to certify the payment due to a contractor. Remeasurement is also known as 'measure-and-value'

Resident Engineer – see *Engineer's Representative*

Retention (retention money) – a part of the payment due to a contractor for progress with the work which is not paid until he has discharged liabilities to remedy defects for a period after the taking over or acceptance of the works by the Promoter. In some contracts this period is called the 'Defects correction period' or 'Defects liability period'. It is typically 12 months

Sanctioning – used in civil engineering to mean deciding to invest in a project, and not used to mean punishing

Schedule of rates – what is called the 'schedule of rates' in some contracts is very similar to a bill of quantities in form and purpose. Contractors when bidding are asked to state rates per unit of items on the basis of indications of possible total quantities in a defined period or within a limit of say ± 15% variation of these quantities. In other examples, the rates are to be the basis of payment for any quantity of an item which is ordered at any time, for instance in term contracts for maintenance and minor construction work and 'dayworks' schedules included in a traditional civil engineering contract

Snagging list – see *Punch list*

Special conditions of contract – conditions added to a set of model or standard conditions of contract to apply to one promoter's projects

Specialist contractors – contractors who limit their work to selected types of work, e.g. piling or building services. They are often subcontractors to general contractors. See also *Trades contractors*

Standard conditions of contract – see *Conditions of contract*

Sub-contractor – a contractor employed by a main contractor to carry out part of a project. A sub-contractor is not in contract with the promoter

Sub-letting – employment of a sub-contractor by a main contractor

Supervisor – used in the new Engineering Contract system to mean the person appointed to check that the Works are constructed in accordance with the contract – see also *Engineer's Representative*

Target-cost contract – a development of the reimbursable type of contract in which promoter and contractor agree at the start a probable 'target' cost for a then uncertain scope of work but also agree that the contractor will share savings in actual cost relative to the target cost but will be reimbursed less than the total extra costs if the target is exceeded

Temporary works – items built to facilitate the construction of the Works

Tender – an offer to enter into a contract, such an offer by a contractor to construct a project. Also called a 'bid'

Term contract – a system in which a Promoter invites several contractors to give prices for typical work which is to be carried out if and when ordered at any time during an agreed period ('the term'), usually based upon descriptions of types of work which may be ordered but without quantities being known in advance

Terms of a contract – all the obligations and rights agreed between the parties, plus any terms implied by law. In English contracts the terms consist of

- *conditions* – the terms which are fundamental to the purposes of the parties to a contract, for instance (i) the Contractor shall construct the Works and (ii) the Promoter shall pay the Contractor
- *warranties* – the less important terms of a contract

Trades contractors – contractors who undertake a class of construction work, for instance electrical installation. Often employed by main contractors as sub-contractors

Turnkey contract – a contract in which the contractor is responsible for the design, supply, construction and commissioning of a complete structure, factory or process plant

Unit rates basis of payment – payment at a fixed price ('rate') per unit of work done. In UK civil engineering 'admeasurement' contracts the predicted total amounts of work are usually stated item by item in a bill of quantities (BoQ) and the contractor is paid the rate for each × actual amount of work done

Valuation – in building and civil engineering contracts in the UK the process of calculating a payment due to the contractor

Value engineering – an analytical technique for questioning whether the scope of a design and the quality of proposed materials will achieve a project's objectives at minimum cost

Variation – a change to the quantity, quality or timing of the works which is ordered by the Promoter's representative under a term of a contract

Warranties – see *Terms of a contract*

Working Rule Agreement – terms of employment agreed between one or more trades unions and representatives of the employers of the members of those unions

The Works – what a contractor has undertaken to provide or do for a promoter – consisting of the work to be carried out, goods, materials and services to be supplied, and the liabilities, obligations and risks to be taken by that contractor. It may not be all of a project, depending upon what is specified in a contract

Works contractor or *work package contractor* – a contractor employed by a management contractor on behalf of a promoter

Works Manager – sometimes the title of a main contractor's General Foreman

For other definitions see Scott J. S. *The dictionary of civil engineering*, Penguin Books, 4th edition, 1991, and in a contract the definitions particular to that contract.

Drafting panel

Professor S. H. Wearne, BSc, DIC, PhD, CEng, FICE, FIMechE, FAPM (Chairman)
Mr W. F. Cotton, OBE, BA, DMS, CEng, FICE, MIHT
Mr M. H. Denley, BSc, CEng, MICE, MIStructE
Eur. Ing. R. W. Giles, OBE, BSc, CEng, FICE, FCIOB, FIHT, FIMgt
Eur. Ing. C. Penny, BSc, CEng, FICE, MIMgt, MIGasE
Professor N. J. Smith, BSc, MSc, PhD, CEng, MICE, MAPM
Mr J. H. A. Thornely, MA, MBA, CEng, MICE, MHKIE, MAPM
Mr T. W. Weddell, BSc, DIC, CEng, FICE, FIStructE, ACIArb
Mrs A. M. Rodgers, BA (Secretary)

The Panel are grateful for the comments and suggestions on the drafts for this new edition given by Eur. Ing. R. L. O. Bennett, BEng, ACGI, CEng, MICE; Mr R. M. Dargie, BSc, CEng, FICE; Mr J. F. Daweswell, MBE, BSc(Eng), FICE, FIHT; Ms M. Ebadi, MEng, GradICE; Mr D. Grimsey, BEng, MSc, MAPM; Mr A. S. Martin, MSc, CEng, FIHT, FBIM; Mr S. Nuttall, BSc(Eng), ACGI; and Miss A. E. Thompson, BA, CompICE.